Making Meaning, Making Money

Making Meaning, Making Money: directions for the arts and cultural industries in The Creative Age

Edited by

Lisa Andersen and Kate Oakley

CAMBRIDGE
SCHOLARS
PUBLISHING

Making Meaning, Making Money: directions for the arts and cultural industries in The Creative Age,
Edited by Lisa Andersen and Kate Oakley

This book first published 2008

Cambridge Scholars Publishing

12 Back Chapman Street, Newcastle upon Tyne, NE6 2XX, UK

British Library Cataloguing in Publication Data
A catalogue record for this book is available from the British Library

ISBN (10): 1-4438-0065-1, ISBN (13): 978-1-4438-0065-5

TABLE OF CONTENTS

ACKNOWLEDGEMENTS

This book would not have been possible without support from the following organisations:

- Regional Arts NSW
- Northern Rivers Writers Centre
- UTS Shopfront Community Program, University of Technology, Sydney
- The Australia Council for the Arts
- Arts NSW

The editors would also like to thank the following people: Paul Ashton and Pauline O'Loughlin from the University of Technology, Sydney, for their advice and guidance; Victoria Keighery, David Throsby, Rachael Vincent, Jeni Caffin, Andrew Griffiths, Lois Randall, Jack Gevorgian, Cathy Henkel, Jeff Canin and Ron Layne for their major contributions to the forums in Byron Bay; Belinda Castles, Pete Fahey, Kim Andersen, Jacqueline Lloyd and Minnie Bannister for their editorial support; and Amanda Millar, Editor at Cambridge Scholars Press, for initiating this volume and her contributions to the manuscript.

Thanks also to all the participants at the Byron Bay forums who help shaped the ideas in this book, especially Bev Abbot, Ien Ang,Neil Armfield, John Birmingham, Julian Burnside, Christine Burton, Stuart Cunningham, Ros Derrett, Nick Earls, Toss Gasgoigne, Helene George, Courtney Gibson, Margaret Fink, John Harding, Clare Keating, Ralph Kerle, Sandy McCutcheon, Jan Muir, Alan Pigott, Scott Rankin, Michael Richards, Rhoda Roberts, Carol Scott, Mandy Thomas, Sam Wagan Watson, Sam Watson and Roger Wilkins.

Special thanks to Jill Eddington for co-creating the forums that led to this publication—and for the lunches.

INTRODUCTION

In his essay for this volume, David Throsby asks whether we are at "a new moment" for cultural policy, and if so, what is the nature of that moment and what does it tell us about the directions that that cultural policy should take us in? The rest of the volume is, in some ways, concerned with trying to answer that question, with authors concurring about the pertinence of this debate, but diverging on much else, including whether future directions for culture can be framed in terms of "cultural policy" alone.

The origins of this collection are in a long lunch—a social formation praised by Tony Moore in this volume for "enhancing creativity"—eaten by Lisa Andersen and Jill Eddington, the then Director of the Byron Bay Writers Festival, not long after Pluto Press's launch of the Australian edition of Richard Florida's *The Rise of the Creative Class*. Feeling the arts and cultural industries had been banished from "father's table" for the decade of John Howard's conservative Australian Government—while still being given pie—the idea that the arts were essential for vital communities was thrilling, as was the knowledge that policy makers in the UK, and elsewhere, were putting the re-branded "creative industries" closer to the centre in decisions about regional development.

Florida's "Creative Class" and the UK's creative industries period were about wealth generation and competitive advantage. But we needed a more integrated explanation of the cultural, economic and social policy dimensions of the arts and cultural industries, alongside building understanding of how individual (or intrinsic) experiences of the arts contribute to personal imagination and public dreaming—all in terms and ways acceptable to a "sector" comprising a broad set of practices and values, suspicious of being trapped in the amber light of conservative, instrumentalist policy. It also seemed important that any (re)imagining of cultural policy should not continue to be left entirely to Academy—though academics have an important role to play—and senior arts agencies, but be directly engaged with the practices and consequences of cultural activity.

From 2004 a series of national arts forums were held at the Byron Bay Writers Festival. Organised by Lisa Andersen and Jill Eddington and supported by Regional Arts NSW, the Northern Rivers Writers' Centre, the Australia Council for the Arts and Arts NSW, the forums brought arts thinkers and practitioners together—out from their chardonnay stocked ivory towers, silos, cultural elites, garrets and siege mentalities—to consider cultural policy, directions for the arts, and ways of developing robust evidence and integrating knowledges and understanding.

The current volume features contributions from a range of people—with diverse perspectives and opinions—who participated in these forums. More than 800 people came, but what was discussed and debated reached a broader audience. This book is the last public outcome from a "product range" that engaged thousands in Australia and elsewhere and included radio broadcasts, webinars, newspaper articles, bibliographies of arts research, audio CDs, podcasts and even a short documentary made by Cathy Henkel and featuring Kate Oakley's keynote speech at the 2004 Forum.[1]

The interest is not surprising. Rarely has culture been so much at the heart of policy discussions in so many places at once. To take but one, important example: China's latest five-year plan is loaded with references to the importance of culture, arguing that, "China not only needs strength in economy, science, technology and defence, but also cultural strength to be ahead of international competition." In that one sentence lies much of the explanation for the current resurgence of interest in cultural policy, the "new moment" as Throsby puts it, but also much that is of concern to traditional cultural policy advocates. Culture has rarely been so much discussed by policymakers; but often only in terms of something else. It has found its way into debates about education and industrial policy, the future of cities and the criminal justice system and, in the form of the creative industries, is increasingly seen as part of international competitiveness, from Shanghai to Sheffield.

Many long-term practitioners and advocates welcome these developments and this embedding of culture in a wide range of policy discussions is evident from the papers collected here. Cathy Henkel's contribution on the Northern Rivers area of New South Wales, is one of several that discuss the role of culture in regional development—in her case the tale of a region stimulated in part through the influx of creative workers and the development of an accommodating milieu. Other papers tell of rural areas from the Highlands of Scotland to Montana where planners—noting that innovations in cultural consumption and creative practices offer opportunities for revitalisation—are looking to repeat this success, albeit in different political and economic conditions. Anne Bamford's focus is on education and on the role that an "arts rich" education can have in developing communication skills, critical thinking and the ability to innovate.

The role of culture within community development, particularly what Martin Mulligan calls "the nature of community in a globalising world," is discussed by several authors—Deborah Mills, Chris Gibson and Paul Brown—as well as Mulligan. Many chapters talk about culture as an aspect of place, whether as the way that communities seek an accommodation within the global market, via tourism for example, or the way they can offer new models of sustainability and perhaps even an alternative to the current dominance (albeit shaky) of neo-liberal economic models.

Phrases like "international competitiveness," "regional development," "education reform," "community cohesion" or "innovation" are thus more likely to crop up in cultural policy documents, and in these pages, than words like "arts" or "beauty," "painting" or "literature." Many of our authors welcome this enlisting of new friends for culture; others are troubled by what these new friends might demand as a price for their company.

In infiltrating these other policy domains, culture has adopted a variety of guises. Most notably, perhaps, it has metamorphosised into "creativity," first as a descriptor for particular industrial activities, the "creative industries," and more recently as an economic necessity; as the forerunner of the innovation on which our economies are said to depend. At the same time, as Chris Gibson notes, creativity is being constructed as a "damaging conflation of socio-economic inequality and cultural attributes," to be deemed "uncreative," is akin to being deemed "feckless" in Victorian times—the difference, for the poor, between being "deserving" and "undeserving."

The new prominence for culture thus raises some concerns, many of which are reflected in this volume. The thread of instrumentalism that runs through much cultural policy discourse worries some writers, not least because there is a danger than in justifying cultural support only in instrumental terms—as a way to reduce crime or boost exam results—there is the chance that another intervention will claim to be equally or more effective, and any sense of the unique claims that culture can make for public support has been lost.

Others argue that despite the lip service (in Anne Bamford's phrase) that policymakers are now paying to all things cultural, the real commitment in terms of funding, support for education, prestige and recognition remains weak. Although "creativity" is lauded, many of the things that feed it—from good classroom experiences to community arts—are being, if not starved, at least put on a restrictive diet.

The tension between rhetoric and reality is also reflected in the clash between growing calls for creativity and what Tony Moore calls the "age of anxiety." At the same time as the economy is said to require creative, critical thinkers and the language of personal development and self-expression is more a feature of contemporary business rhetoric than it is of contemporary politics. It is important to examine what is meant by "creativity" in the rhetoric and to ask how it can thrive in atmospheres of fear, constraints on civil liberties, economic downturn and precarious employment.

Culture would at one time have been seen as one of the natural homes for discussing such issues—a time honoured space for social critique and self-examination, even in a hostile environment. Tony Moore's contrast between the response of artists in Thatcherite Britain and the response of artists to today's anxieties deals with this issue directly. One relevant difference may be that, although culture was almost certainly undervalued by the Thatcher government, that very marginalisation gave it a space in which to operate. Today's practitioners often have to tread carefully to ensure that being uncritically embraced for the magic dust they can sprinkle on the economy does not restrict their freedom of movement in other spheres.

None of our authors suggest that developing a national cultural policy would resolve all, or even most, of these issues; though, as Christopher Madden argues, it might help to develop a logical coherence for government's various involvements in culture. More importantly, the process of developing a policy might open up a space for debate that is, in itself, as important as anything a policy document will contain. And even those of our authors who are sceptical about the value of a national cultural policy—or even our capacity to develop one—have ideas about what debates it needs to explore.

David Throsby is clear that no debate about cultural policy can start without considering the benefits of the subsidised arts system and, within that, debating what funding priorities should be. He makes a strong case both for community cultural development and for new media; the first because of its broad social contribution, the second because of its links to wider economic innovation. Deborah Mills and Paul Brown both make eloquent cases for the role of community cultural development, which Brown sees as a constituting "a site for the production of important knowledge about complex social, economic and environmental matters," and which Mills sees as a way of getting beyond the rather crude instrumentalism of much "arts as social policy approaches" and instead as way of embedding "cultural sensibility in the ways that governments think and act." Or as Chris Gibson argues, in his chapter on regional development, we "need creativity in policy-making much more than overt policy prescriptions about creativity."

That cultural policy should indeed touch on the role of culture within the economy and on the "creative industries" agenda is generally agreed, though there are divergences on how and to what extent this needs to happen. It is the sustainability of some of the claims for the creative industries and the competitive framework into which they are often placed that concerns us. As Chris Gibson argues: "What seems to be happening is that a singular interpretation of creativity is being incorporated into a rather uncreative framework, in which private sector solutions to regional problems, the ideal of creative, independent, entrepreneurial subjects, and the primacy of place competition in global markets remain paramount." And as Cathy Henkel, in what is generally a positive picture of region that has benefited from an influx of creative practitioners, points out, success brings it own problems in terms of place development, gentrification, higher prices and marginalisation of both lower-paid practitioners and other low-income residents. The current economic downturn adds still further questions to a model based on growing consumer markets.

Overcoming this, as Chris Gibson suggests above, is as much about the institutional structures and administrative arrangements through which policy is delivered, as it is about policy development. As Deborah Mills argues, cultural policy must be sensitive and aware of the cultural consequences of all government policies and activities—a sort of "culture proofing" of other public policies. David Throsby contends that extending the ambit of cultural policy in this way, to embrace the cultural underpinnings of other policies, would be something relatively new in

Australia but would open up a variety of possibilities and questions about issues as diverse as the Australian republic, the treatment of refugees and indigenous affairs.

It is clear that the embedding of culture within wider policy frameworks can only increase the calls for improved evidence as to the effectiveness of these collisions. As researchers, consultants, activists, academics and practitioners, most of our authors are sympathetic to calls for "more research," though again the nature of much of what passes for evidence is questioned. As Paul Brown points out, evidence-based processes are deemed important by governments, yet much of the evidence gathering is structured "from the centre," reflecting funding programmes and temporary policy concerns, rather than a deep desire to understand the complex ways in which culture interacts with wider society. And as Kate Oakley points out, on the rare occasions that serious research is carried out on cultural impacts, policymakers and even arts advocates are often disappointed by the results—there are not as many people "saved" from crime, not as many sustainable jobs, as some of the more hyperbolic rhetoric would lead you to believe.

The landslide election victory for Kevin Rudd's Labor government in Australia in late 2007 was greeted with enthusiasm by many in the cultural sector alongside a sense of relief that new ideas could finally get into circulation—including some old new ideas. Almost immediately The Prime Minister's Literary Award—one of the richest in the world—was announced. Then, in February 2008, Rudd delivered the long-awaited apology to the Stolen Generations of Indigenous Australians at the opening of Federal Parliament, followed, in April, by the inclusion of "Creative Australia" as a critical area of policy for discussion of the 2020 Ideas Summit, with accompanying star power from Cate Blanchett and Hugh Jackman.

And more star power was injected into Creative Australia in the form of millionaire, rock icon, Peter Garrett, as the first Australian Minister for the Arts drawn from a successful career in the arts, who, as Tony Moore says, "knows the value of risk-taking." His first arts policy, *New Directions in the Arts* (2007), included creative industry concepts of entrepreneurialism and innovation alongside greater support for arts education, community engagement, equity and the "democratisation" of the arts, with a focus on Indigenous arts and young and emerging artists.

While Indigenous artists and visual artists welcomed the resale royalties scheme, the Government's first budget contained no increase in spending—the slice of pie is the same. And in May the cultural sector got a look at Kevin Rudd's position on artistic freedom when he condemned an exhibition by photographer Bill Henson containing images of nude children as "revolting." So, as David Throsby says, it is "too early to tell whether the Rudd Government will deliver a new dawning for the arts and culture in Australia—whether in fact a new moment for cultural policy has arrived." But the signs for the eternally hopeful cultural sector are that some politicians *have* been listening to the ideas and the debates you will find in this volume—and the policy landscape on "creative" *is* moving away from "new markets." In his address to the National Press Club on 3 September, 2008, the Innovation Minister, Senator Kim Carr, said:

> I believe the creative arts—and the humanities and the social sciences—make a terrible mistake when they claim support on the basis of their commercial value.
>
> Whatever they may be worth in the marketplace, it is their intrinsic value we should treasure them for.
>
> We should support these disciplines because they give us pleasure, knowledge, meaning, and inspiration.
>
> No other pay-off is required.

---

[1] Available from www.regionalartsnsw.com.au

PART I:

THE CULTURAL POLICY MOMENT

CHAPTER ONE

A NEW "MOMENT" FOR CULTURAL POLICY?

DAVID THROSBY

Introduction

In August 1994 a special issue of the journal *Media Information Australia* appeared, edited by Terry Flew, Gay Hawkins and Elizabeth Jacka. The title of the issue was "The Policy Moment," and it contained a collection of articles on contemporary issues in cultural policy and cultural studies by a wide range of academics, bureaucrats and policy-makers. In an overview essay at the front of the collection, Stuart Cunningham summed up the state of affairs as follows:

> Historically speaking, we seem to be living in a cultural policy "moment." In Australia as in many countries…cultural policy has hitherto been a backwater—dominated by constricting assumptions about "high" art and worthy subsidy; narrow in its consideration only of the arts rather than of culture more generally (an aesthetic rather than an anthropological concept of culture) (Cunningham 1994, 4).

In contrast, Cunningham went on to argue, policy-makers in the early 1990s were increasingly looking to the notion of culture as embodying a framework for livability in conditions of economic, social and environmental change. He suggested that the cultural policy "moment" had arrived. He noted that

> [b]oth the anthropological/institutional notion of culture (and its offspring the deployment of culture industries) and the broad rather than narrow definition of culture as a whole way of life are necessary preconditions for the emergence of the moment. Each of these preconditions now has a foothold in Australian public policy discourse at all levels of government (1994, 5).In contrast, Cunningham went on to argue, policy-makers in the early 1990s were increasingly looking to the notion of culture as embodying a framework for livability in conditions of economic, social and environmental change. He suggested

that the cultural policy "moment" had arrived. He noted that

> [b]oth the anthropological/institutional notion of culture (and its offspring the deployment of culture industries) and the broad rather than narrow definition of culture as a whole way of life are necessary preconditions for the emergence of the moment. Each of these preconditions now has a foothold in Australian public policy discourse at all levels of government (1994, 5).

In the same month in which this up-beat assessment appeared, the Federal Government's Department of Communications and the Arts organised a two-day forum in Canberra under the title "Creative Culture: the New Growth Industries" at which a comprehensive gathering of researchers, government officials and representatives of arts and cultural organisations talked in glowing terms about the expanding prospects for the creative industries in the Australian economy and about how government policy could facilitate their growth (Department of Communications and the Arts, 1994). Several months after this forum, in October 1994, the government's cultural policy statement *Creative Nation* was released. The cultural policy moment had arrived indeed.

Since 1994 there have been two changes of Australia's federal government, both of which have had significant implications for cultural policy. The first was the replacement in 1996 of Paul Keating's progressive Labor government by John Howard's Liberal/National coalition leading to a turn towards a more conservative stance in cultural matters in the years that followed. Then, after eleven years of the Howard administration, the political wheel revolved once more, bringing about the election in November 2007 of the new-look Labor Party led by Kevin Rudd. The enthusiasm with which the Australian people disposed of the old and embraced the new at the 2007 election might be taken to herald the possibility of a more progressive direction in cultural policy, prompting the question: has a new cultural policy moment come upon us, strangely reminiscent of the one that occurred almost fifteen years ago?

In this paper I discuss this question, picking up on some of the themes developed in my *Platform Papers* essay "Does Australia Need a Cultural Policy?" (Throsby 2006), and extending some of those themes further in the light of discussion generated since that essay's publication.[1] I begin in section 2 by looking briefly at what cultural policy is and how it grew up in Australia, culminating (in a sense) in the appearance of *Creative Nation*, discussed in section 3. I then consider in section 4 some areas of cultural change that emerged in Australian society over the Howard years. In section 5 I discuss developments since the 2007 election, and conclude with some observations about the future direction of cultural policy in this country.

2. The ambit of cultural policy

In most fields that are of interest to government policy such as the environment, health, education, transport and so on, the area of concern is clearly identified, and hence the scope of policy coverage is reasonably clearly delineated. Not so with culture. There are many questions surrounding the definition and usage of the word "culture" and hence of the related concept of "cultural policy." Does the term "culture" embrace only what are commonly referred to as the high arts—drama, dance, opera, "serious music," literature, the visual arts? Or does it extend to popular culture, including film, video games, television, rock music and other similar fields? Or is an anthropological or a sociological definition of culture more appropriate, as alluded to by Cunningham above, extending the coverage beyond creative activities and focussing attention on matters of identity, values, beliefs, tradition, ways of living together—in short just about anything that serves to identify a group and bind it together? It is apparent that what comprises cultural policy will vary widely according to which interpretation of culture is used.

In practice, however, there is little doubt that historically in Australia the term cultural policy, to the extent that it has been used at all, has meant arts policy. So, for example, a document entitled *Cultural Policy in Australia* written in 1980 by Jean Battersby, the first administrative head of the Australia Council, was principally concerned with the high arts as defined above, although she did include chapters on film and broadcasting, and there was even a section on sport (Battersby 1980). Likewise, twenty years later, Deborah Stevenson's book subtitled *Making Australian Cultural Policy* was also concerned primarily with government policy towards the arts, although her orientation as she charted the changing environment over the last couple of decades was anything but elitist (Stevenson 2000).

The development of Commonwealth arts policy since Harold Holt moved to set up the first Australian Council for the Arts in 1967 has been documented by a number of writers.[2] The story is briefly summarised as follows. The first half of the 1970s saw a substantial consolidation and expansion of arts support through the establishment of the Australia Council by the newly-elected Whitlam Government. Soon after it was born, the Council had to endure a searching review of the rationale for its functions carried out by the Industries Assistance Commission (1976), and then faced another enquiry ten years later by a House of Representatives Standing Committee on Expenditure (1986), chaired by Leo McLeay. Despite their critical demeanour, neither of these enquiries brought about any significant changes in the Australia Council's operations, nor did they upset the willingness of the government of the day (Coalition in the first case, Labor in the second) to continue funding the arts from the public purse. Indeed by compelling arts advocates to justify their position, these enquiries helped to consolidate in a local context the economic arguments for government intervention in this area, arguments which were being well worked over by economists elsewhere.[3]

The articulation of an economic basis for arts policy in terms of public goods provision during the 1980s was given an added dimension by the climate of economic liberalisation and microeconomic reform prevailing at that time. In this environment lobbyists found it advantageous to promote the arts as generators of economic wealth, arguing that the arts made significant contributions to incomes, exports, employment and regional development. Such a characterisation of the arts—as an industry producing goods and services for sale to consumers rather than as a sublime expression of human creativity—ruffled many feathers at the time, and still does. Critics such as Donald Horne[4] saw these developments as implying a relentless commodification of the arts and a sell-out to the soulless rigour of the marketplace. In fact, labelling the arts as an industry does no such thing; it simply acknowledges the fact that artistic production and consumption occur in an economic context which can be analysed using the tools routinely employed by economists to study industrial organisation, consumer behaviour and market disequilibrium. Doing so does not deny that the role of art in individual experience or in society at large serves a much larger purpose.

Despite its critics, the concept of the cultural industries that started to take shape during the late 1980s reached its apotheosis in the Keating Government's grand vision for a cultural policy referred to above. But *Creative Nation* was much more than a blueprint for industry development. It represented the first serious effort to put together a comprehensive national cultural policy, springing from the very roots of what could be described as indigenous and non-indigenous Australian culture. As such, it represented a decisive shift away from the narrowness of earlier policy statements concerned primarily or exclusively with support for the arts.

3. Creative Nation

Given that *Creative Nation*[5] was the first attempt to spell out in a single document what an Australian national-level cultural policy could look like—what Christopher Madden refers to elsewhere in this volume as a "monolithic cultural policy" —it is worth dwelling on how it was framed and what it prescribed in terms of actual policy implementation. It is important at the outset to note that *Creative Nation* was motivated overwhelmingly by ideas about Australian identity and cultural sovereignty. The very first paragraph of the Introduction reads:

> To speak of Australian culture is to recognise our common heritage. It is to say that we share ideas, values, sentiments and traditions, and that we see in all the various manifestations of these what it means to be Australian (p.15).

The ways in which a national-level cultural policy would reflect these sentiments would be

> by shoring up our heritage in new or expanded national institutions and adapting technology to its preservation and dissemination, by creating new avenues for artistic and intellectual growth and expression and by supporting our artists and writers (p.7).

By these means Australia would "ride the wave of global change in a way that safeguards and promotes our national culture" (p. 7).

As noted above, *Creative Nation* saw itself as being not only a cultural policy, but also an economic policy. To some extent this picked up on the 1980s trend towards interpreting arts policy in economic terms to which I drew attention earlier. But it also extended these arguments much further, by embracing the cultural ramifications of new communications technologies, first through an emphasis on new media as means for cultural production, participation, dissemination and access, and second through a broader focus on cultural industry development. The essential element common to both cultural and economic progress was seen to be creativity; on the one side artists would generate innovative ideas reflecting our distinctive cultural identity, while on the other side creativity would be fostered as a key resource in the development of the new economy.

The report saw the Federal Government's role in cultural development as comprising a number of strategies including: nurturing creativity and excellence; enabling all Australians to enjoy the widest possible range of cultural experience; celebrating Indigenous art and culture; preserving Australia's heritage; promoting the expression of Australia's cultural identity, including its great diversity; and developing lively and sustainable cultural industries, including those evolving with the emergence of new technologies. The authors of the report proposed a variety of measures to achieve these ends, including: an expansion of the Australia Council; a range of specific initiatives in the performing arts, the visual arts and literature; increased support for film and public broadcasting; new media development; various industry strategies; expansion of the Commonwealth's role in the protection and preservation of cultural heritage; and other programs in education and training, market development and tourism. In short, *Creative Nation* provided not only a rationale and advocacy for an active Commonwealth role in Australia's cultural development, but also a set of proposals with resources attached for putting that role into effect.

For its time, *Creative Nation* was an exceptionally forward-looking document, anticipating several lines of cultural policy development which have since found expression in other countries, if not here in Australia. For example, its efforts to link the cultural industries with new media technologies and the growth of a creative workforce prefigured later moves in these directions in the UK, the US and other countries.[6] Nevertheless, despite its heroic intent, *Creative Nation* proved to be longer on rhetoric than on any real capacity to deliver cohesive and operational measures to bring about change. True, it gave rise to some successful new programs in the cultural field, but it also produced some spectacular failures. Overall, it did not lead to any dramatic shifts in cultural policy formulation or implementation at the national level. Any chance that, given time, it might have yielded longer term fruits was effectively scuttled by the change of federal government in 1996. Curiously, the overall spirit of *Creative Nation* has become, in the ensuing years, more apparent in the States than at the Commonwealth level, insofar as it has been the States rather than the Commonwealth that have picked up on the creative industries agenda. Most of the States have shown interest in industry development strategies in which new media, digital content technologies and the creative industries more generally play a significant role, especially as components of initiatives for urban or regional growth.

4. Cultural change 1996 to 2007

How did Australian society change in cultural terms during the eleven years of conservative rule? Cultural change is a gradual process affected by a variety of internal and external influences. Certain cultural values may seem long-lasting and resistant to change, others may be more fluid; either way measuring cultural change is problematical because culture itself is not a single fixed entity but a mix of many different components. Furthermore at a policy level it is unclear to what extent cultural change occurs autonomously within society (as a result, say, of demographic shifts) and to what extent it is influenced by government action. And if governments do indeed influence the directions of cultural change, are they behaving proactively in shaping cultural values, or simply reacting to what they perceive to be the mood of the people?

Looking back over the Howard years we can identify both external and internal influences that affected Australian culture in one way or another. The major external factors have been the inexorable processes of globalisation, and the rise in global insecurity following 9/11 and the invasion of Iraq. In regard to the former, debate on the cultural impacts of globalisation continues without, so far, any decisive result—evidence can be found both for the penetration of universal cultural symbols (mostly of U.S. origin) on the one hand, and for the resilience of local cultural identities on the other. The cultural ramifications arising from the fear of terrorism are more disturbing. Incumbent governments in Western countries have been able to draw some political advantage from a heightened sense of insecurity amongst the population, aware that in times of uncertainty people tend to stick with the government they know. More broadly, a process of cultural polarisation, crudely summarised in the title of Benjamin Barber's 1996 book *Jihad vs. McWorld*, has been given added impetus by contemporary terrorism, placing the ideal of the multicultural society under considerable strain.

At the same time as these external influences were having some effect on cultural change in Australia, internal factors also played their part. The coming-of-age of the baby-boomer generation and continuing high levels of economic growth had produced a comfortable middle-class Australian society that seemed quite content with the way things were. The traditional cultural concept of the "Aussie battler" was replaced with something more affluent. Whether he or she was happier as a result remains an open question, as does the matter of what comprises Australian identity in the post-millennium world.

Given these various influences, what can we say about the extent to which the former Federal Government and/or the Prime Minister himself influenced processes of cultural change in Australia over the eleven-year period? Alternatively, to what extent did the explicit or implicit cultural positions adopted by John Howard and his Ministers coincide with, or were out of step with, the cultural mood of the Australian people? This is a large and complex question deserving of an entire volume. Here I consider just three areas, conveniently tagged as the three Rs: reconciliation , refugees and the republic.[7]

First, there can be little argument that the process of reconciliation between non-Indigenous and Indigenous Australians that had been gathering momentum during the early 1990s was effectively stalled in 1996. While expressing concern about the appalling state of Indigenous health, education and welfare, the Federal Government under John Howard showed little understanding of fundamental issues affecting the relationship between European and Aboriginal society—the importance of land rights, the significance of an apology, the centrality of Indigenous culture, Aboriginal people's need for respect rather than condescension. The sentiments expressed in Paul Keating's celebrated Redfern speech of December 1992, or in the dignified and heartfelt apology written on John Howard's behalf by John Clarke and delivered by the PM's actor namesake in the run-up to the 2000 Olympic Games,[8] are not likely to have been shared by the man himself. In view of the Australian community's apparent sympathy for the idea of reconciliation, as evidenced by a number of public demonstrations of support for Indigenous rights, it can be concluded that the former Prime Minister's position in this area was out of harmony with the prevailing popular sentiment.

Second, the Coalition Government's policy towards the treatment of refugees in detention continued to reflect cultural attitudes that would appear to an outside observer to be in marked contrast to traditional Australian values of fair play and respect for human rights. Some have argued that the results of the "Tampa" election of 2001 vindicated the government's stance on refugees; if so, this would indicate a significant success for the government in shifting Australian cultural attitudes towards a less easy-going, less tolerant and more individualistically-centred society. However, given the range of other factors that affected the outcome of that election, it is dangerous to draw such a simplistic conclusion. It seems more likely that the fundamental Australian cultural value of the "fair go," though by now irrevocably cliché-ed, is more enduring and hence that the ways in which the former Federal Government dealt with refugees was inconsistent with basic cultural attitudes in this country.

Third, the period after the Second World War saw the process of Australia's emergence at last from its colonial shadow gather momentum, a process hastened by Britain's application to join the European Common Market in the 1960s. During the 1980s and early 1990s the self-confidence of Australia as an independent nation in the Asia-Pacific region grew, consolidating a sense of Australian culture as something unique and worth protecting. The distinctiveness of Australian cultural identity continued to be celebrated with some exuberance at festivals such as the opening of the Sydney Olympic Games in 2000 and of the Melbourne Commonwealth Games in 2006. Nevertheless at a broader level our sense of cultural independence and self-determination seemed to be called into question by several of John Howard's policy positions. For example, by shrewd political manipulation he effectively scuttled the republican cause in the 1999 referendum and in so doing he ensured that the republic was off the agenda for the remainder of his term. Furthermore, he moved much closer than his predecessors in aligning Australia's foreign policy interests with those of the United States, for example by committing Australia to war in Iraq when a clear majority of the population was opposed to such action. The willingness to please our great and powerful ally had significant cultural ramifications, seen, for example, in our failure to exclude culture from the US/Australia Free Trade Agreement negotiated in 2003–04 and in our conspicuous abstention in the vote for adopting the UNESCO Cultural Diversity Convention in October 2005. Both these actions could be read in cultural terms as a meek compliance with pressure for Australia to toe the American line, and raised questions about the path of our cultural development in the future.

5. Change and challenge post-2007

During the election campaign in November 2007, John Howard portrayed the choice facing the Australian people as one between political correctness and traditional values. He went so far as to claim that his conservative stance on cultural issues was one of the most important distinguishing characteristics of his period in office, and he warned that election of a Labor government would have disastrous consequences for Australian cultural life. Predictions of such a calamity made little impression on the electorate; indeed there is evidence that a cultural shift was already underway that had turned against the sorts of values that Howard and his ministry represented. Both the Prime Minister himself and the government he led were soundly defeated at the election, and a Labor government was duly installed.

It is too early to tell whether the Rudd Government will deliver a new dawning for the arts and culture in Australia—whether in fact a new moment for cultural policy has arrived. So far (by mid-2008) the signals have on the whole been positive, although not entirely so. Certainly in regard to two of the three Rs noted above—reconciliation and the republic—progress has been made. The incorporation of Aboriginal ritual into the opening of the new Federal Parliament in February 2008, and the apology to the Stolen Generations delivered on that occasion, touched the hearts of Australian people in a truly remarkable way, and did much to rehabilitate Australia's cultural reputation in the world at large. Similarly, the Prime Minister has placed the republican issue back on the table, with an observation that it is now a matter not of "whether" but of "when" we will have an Australian as Head of State, although no time-table has yet been set for carrying this forward. . In regard to refugees, some of the more deplorable aspects of the former government's policies have been overturned, but there is much still to do. .

In the broader cultural policy domain, the most significant event since the election of the new Federal Government has been the inclusion of a "Creative Australia" stream as one of the ten policy areas for discussion at the 2020 Ideas Summit held in Canberra in April 2008. This stream brought together one hundred artists, academics, administrators, media people and others who talked for two days about the long-term development of our culture and put forward ideas for making creativity a central element in our lives and our economy. The resulting recommendations[9] provide a blueprint for action across areas such as support for the arts, sustainability of arts practice, broadcasting and screen culture, the development of Indigenous art, arts education, architecture and design, cultural diplomacy, and many more.

The palpable enthusiasm with which Kevin Rudd was received by arts participants in the 2020 Summit soured somewhat subsequently when, in a controversy over censorship of artworks, he adopted what could be read as an extremely populist position. Just when there was hope that the worst of the culture wars was behind us, the flames flared again. It is unclear whether the Prime Minister's interventions in this debate indicate a fundamental conservatism in his cultural outlook. Perhaps it does, although his positive remarks about the arts in his concluding speech to the Summit might suggest otherwise. Whatever the truth of the matter, there are many hopeful signs that the new Government's overall cultural policy will continue to move in progressive directions.

6. The way ahead

In my *Platform Papers* essay I advocated a "bottom-up" approach to cultural policy, whereby grass-roots discussions about our arts and culture could feed into a broader national debate. The 2020 Summit has clearly been a start in this direction, but the question remains as to how such a debate can coalesce into an overall cultural policy or into a series of specific policies with cultural content. In concluding this paper I point to five areas where policy attention could be directed. This is not intended to be a comprehensive list of required policy measures, simply a selection of some important areas for particular consideration.

First, we need to make a critical assessment of where our current arts policy is getting it right and where it is falling short. A first step would be to undertake a comprehensive look at the wider benefits of the arts. If past empirical evidence is anything to go by, an increase in current levels of Federal Government support for the arts would not be out of line with consumer preferences. More generous funding could catalyse a new renaissance in artistic production and access across the board. But it is not just a matter of more money, it is as much a question of how the money is used. It is ironic that two significant areas where policy attention is needed are precisely those areas whose boards have been abandoned by the Australia Council—new media and community cultural development. Although the Council has insisted that its new arrangements are continuing to cater for these areas, the symbolic implications of the changes have been unfortunate. Both of these areas have a great deal of relevance in the present cultural circumstances: community cultural development because of its positive role in generating social cohesion and understanding in ethnically diverse communities, including Indigenous communities, and new media because they are essential in making the arts' contribution to the information economy happen.[10]

This leads to a second policy arca, the creative industries agenda. Much hype surrounds claims concerning the economic contribution of the cultural industries, as Kate Oakley's paper in this volume points out. Nevertheless, in Australia there is accumulating evidence that, properly managed, these industries can contribute to sustainable growth, especially in regional areas where particular creative concentrations can be identified and fostered. Cultural policy in this field needs to focus on creativity and how the arts, as the core element of the cultural industries, can contribute to creative dynamism in the wider cultural industries as well as in other sectors of the economy. A step towards reclaiming a Commonwealth role in the development of our cultural industries is provided by the Federal Government's provision of funds over three years from 2008 for a Creative Industries Innovation Centre, a network to deliver business services to small-to-medium enterprises in the creative sector across Australia.

Thirdly, policy analysis should always pay attention to the institutional structures and administrative arrangements through which policy is delivered. In the cultural policy arena this means a wide array of cultural institutions, from public museums and galleries to the ABC and SBS. It also includes funding bodies such as the Australia Council and the newly-formed Screen Australia. An expansionary view of cultural policy in Australia would see some relaxation of the financial constraints under which all of these institutions, almost without exception, are labouring. Again, however, it may not be simply more money but also a review and, if necessary, re-affirmation of their aims and structures and the principles on which they are founded.

Fourthly, extending the ambit of cultural policy to embrace the cultural underpinnings of policies in other areas would be something relatively new in Australia and yet would be consistent with ideas of cultural openness and consistency in policy formation at the national or State level. For example, we might ask questions such as: What are the cultural implications of economic and social policies in the field of Indigenous affairs? How will changes to cross-media ownership affect Australian cultural life? How can we give more recognition to the non-monetary value of cultural goods and services in up-coming trade negotiations? These are questions that should be asked as a normal part of the process of policy development. They could be addressed if policies were required to submit to a cultural impact assessment, in the same way that environmental impact assessments are required for development projects. This is an idea that was mooted in the early 1990s; it could well be exhumed for reconsideration now.[11]

Finally, all aspects of cultural life—creation, participation, enjoyment, understanding—depend on an educated and well-informed population. Education, meaning in this context everything from pre-school child care to life-long learning, is the foundation stone upon which a flourishing culture is built. Any cultural policy must not fail to recognise the importance of education in the arts particularly, and in the humanities more generally, in building a civilised society. The recommendations flowing from the 2020 Summit in regard to education in the arts provide a basis for doing this; they deserve to be given priority attention.

7. Conclusion

To conclude I refer again to Stuart Cunningham's essay which I quoted at the start of this paper. In responding to critics who suggested that the cultural policy moment of the early 1990s was simply a reflection of the hubris of Labor in power, he ventured the opinion that "there is little reason to expect that engagement with cultural policy would become untenable under a conservative government in Australia" (Cunningham 1994, 5). No doubt his optimism was founded on the historical evidence—conservative governments of the past have been favourably disposed towards the arts and culture, and the development of policy initiatives in these areas has been by no means a Labor prerogative. And indeed John Howard, in reviewing his period in office, defended his government's record in the arts; in an interview with the *Bulletin* on February 20, 2006 Mr. Howard said, "we spend more money on the arts then we do on sport and that's even leaving aside the ABC… Now people will immediately say it's easier to get sports sponsorship than the arts. That may be right. It doesn't alter the fact that our financial support for the arts is actually greater than for sport."

Two observations can be made on the former Prime Minister's comment. First, it is significant that he chose sport as the point of comparison, as if the fact that spending more on the arts than on sport made it all alright. Second, his remarks were made in response to the question "Would you see at any time a Howard government looking at [a cultural policy]?" Mr. Howard's answer implied that for him a cultural policy continued to be simply an arts policy and nothing more.

The cultural landscape has now changed. Notwithstanding some hiccups, so far the outlook is promising for a clearer comprehension by the Federal Government of what cultural policy is, and in what directions it should be moving. I suggest that the key word in Stuart Cunningham's prediction quoted above is "engagement." There is hope that the new cultural policy moment in this country is indeed one of heightened engagement by governments and by the people in fostering our artistic and cultural development in the future.

References

Barber, Benjamin. 1996. *Jihad vs. McWorld: how globalism and tribalism are shaping the world.* New York: Ballantine Books.

Battersby, Jean. 1980. *Cultural policy in Australia.* Paris: UNESCO.

Commonwealth of Australia. 1994. *Creative nation: Commonwealth cultural policy.* Canberra: Department of Communications and the Arts.

—. 2008. Towards a creative Australia: the future of the arts, film and design. *Australia 2020 Summit: final report.* Canberra: Department of Prime Minister and Cabinet, pp.253–301. Available at http://www.australia2020.gov.au/final_report/index.cfm, accessed 21 July 2008.

Council of Europe and ERICarts. 2002. *Cultural policies in Europe: a compendium of basic facts and trends*, Strasbourg: Council of Europe.

Cunningham, Stuart. 1994. Willing wonkers at the policy factories. *Media Information Australia* 73: 4–7.

Department of Communication and the Arts. 1994. *Creating culture: the new growth industries.* Canberra: Australian Government Publishing Service.

Flew, Terry. 2002. *New media: an introduction.* Melbourne: Oxford University Press.

Florida, Richard. 2002. *The rise of the creative class: and how it's transforming work, leisure, community and everyday life.* New York: Basic Books.

Gibson, Lisanne. 2001. *The uses of art: constructing Australian identities.* St. Lucia: University of Queensland Press.

Horne, Donald. 2001. *Looking for leadership: Australia in the Howard years.* Ringwood: Viking.

House of Representatives Standing Committee on Expenditure. 1986.

Patronage, power and the muse: inquiry into Commonwealth assistance to the arts. Canberra: Parliament of the Commonwealth of Australia.

Industries Assistance Commission (1976), *Assistance to the Performing Arts*, Canberra: Australian Government Printing Service.

Macdonnell, Justin. 1992. *Arts, Minister? Government policy and the arts.* Sydney: Currency Press.

Parsons, Philip, ed. 1987. *Shooting the pianist: the role of government in the arts*. Sydney: Currency Press.

Rowse, Tim. 1985. *Arguing the arts: the funding of the arts in Australia.* Ringwood: Penguin.

Selwood, Sara, ed. 2001. *The UK cultural sector: profile and policy issues*. London: Policy Studies Institute.

Smith, Chris. 1998. *Creative Britain*. London: Faber and Faber.

Stevenson, Deborah. 2000. *Art and organisation: making Australian cultural policy*. St Lucia: University of Queensland Press.

Throsby, David. 2001. Public Funding of the Arts in Australia: 1900-2000. *Year Book Australia 2001*. Canberra: Australian Bureau of Statistics, 548–561.

—. 2003. Does the Australian Government Have a Cultural Policy? *Dialogue* 22 (2/2003) Academy of the Social Sciences in Australia, 54–60.

—. 2006. Does Australia Need a Cultural Policy? *Platform Papers* 7: 1–54.

Towse, Ruth, ed. 1997. *Cultural economics: the arts, the heritage and the media industries*. Cheltenham: Edward Elgar, 501–719.

United Nations Conference on Trade and Development (UNCTAD) 2008. *Creative Economy Report 2008*. Geneva: UNCTAD.

Notes

[1] The present paper draws to some extent on that essay (a very early version of which was presented at the Byron Bay forum from which this volume derives) and also on an earlier paper written for the Academy of Social Sciences in Australia (Throsby 2003).

[2] For overviews of the development of arts policy in Australia see, for example, Rowse (1985); Parsons (1987); Macdonnell (1992); Stevenson (2000); Throsby (2001); and Gibson (2001).

[3] These arguments revolve mainly around the possibility of the arts being a case of market failure, where government intervention is needed to provide public goods

or to enhance the output of beneficial externalities; for an overview, see papcrs collected in Towse (1997).

[4] See, for example, Horne (2001), and his address to Sydney Arts Management Advisory Group, Australia Council, Sydney, 24 March, 2003.

[5] See Commonwealth of Australia (1994).

[6] See, for example, Smith (1998); Selwood (2001); Florida (2002); Council of Europe and ERICarts (2002); and UNCTAD (2008).

[7] This is two fewer than Frank Brennan's five Rs in his 2006 Manning Clark lecture: race, religion, respect, rights and the republic, see http://www.manningclark.org.au/papers/MCLecture-2006.html (accessed 21 March 2006).

[8] Reproduced as an appendix in Horne (2001) p. 285.

[9] See Commonwealth of Australia (2008).

[10] For a view of the role of new media across a wide range of areas in culture and the arts, see Flew (2002).

[11] Here is a small example. When roads in rural areas are widened or new roads planned, they sometimes impinge on Aboriginal land or affect sacred sites. Several State governments have adopted procedures requiring consultation with local Aboriginal communities about the cultural effects of road projects in country areas, to ensure that the Indigenous cultural values would be respected in decisions relating to such projects.

CHAPTER TWO

ANY ANSWER AS LONG AS IT'S RIGHT: EVIDENCE-BASED CULTURAL POLICYMAKING

KATE OAKLEY

"There is no objectively correct way to govern.

Some public policy issues involve trade offs between competing resources—these will always be a matter for the values of political parties and their leaders."
UK Strategy Unit, February 2005

"If someone says evidence-based policy making one more time to me, I'll deck them."
Louise Casey, Head of Anti-Social Behaviour Unit, July 2005

Introduction

This is a story about UK creative industries policy as it has developed under the "New Labour" administration. How it has been shaped, influenced and in some cases hijacked by the competing goals of economic growth, urban revitalisation and poverty reduction. How it has been de-coupled from "culture" and then re-linked to it. And it is a story about "evidence" and the limits to evidence, especially when faced with a Government whose stated commitment is, "what matters is what works." Forgetting of course that the real question of politics is always; works for whom?

But it is not a parochial story. The debate about "creative industries" and their relationship to the arts and the broader cultural realm has been more self-consciously played out in the UK than in many other countries—in part because the term was coined there and in part because these activities really matter in the context of the UK's economy. But there are echoes elsewhere. The debate in the US fired by Florida's *Rise of the Creative Class* (2002), also touched on issues of evidence, on the instrumentality of arts and cultural policies and on the suppression of the conflicts inherent in "culturally-led regeneration." (Siegel 2005; Kotkin 2005) Canadians have to some extent defined their own urban cultural policies in contrast to this market-led gentrification, (Gertler 2004); while Gibson and Klocker (2005), have examined how a particular view of "creativity," is being used to drive place-based competition in rural and regional Australia.

Despite these echoes, the stated enthusiasm for evidence-based policymaking is perhaps nowhere as intense as in the UK. The first quote above is taken from a document entitled, "Strategic Audit, Progress and Challenges for the UK," which was released by the UK Cabinet Office in February 2005. The document is a policy-wonk's dream, full of charts and graphs about employment and inflation, skills and R&D, health, crime and environmental impacts. It is rigorous and informative—comparing the UK's use of renewable energy sources to that of Sweden (one up to Sweden) or international trends in social trust (Swedes win again). As an aid to policy making it may well be invaluable.

However, page six of the document contains the rather odd statement displayed above,

> There is no objectively correct way to govern. Some public policy issues involve trade offs between competing resources—these will always be a matter for the values of political parties and their leaders.

So that's all right then. Just in case you were running away with the idea that the good society was simply a matter of comparing spreadsheets, of totting up the numbers and deciding who was on top and that all the values, tension and competing ideas had been removed from the political landscape, the authors of the report are there to reassure you that politics still contains some ideas to be fought for and debated and some conflicts that may never be resolved into a win-win consensus.

The fact that such a notion needs stating at all suggests that we had perhaps reached the high-water mark of commitment to evidence-based policymaking (EBP), somewhere in the early spring of 2005. Indeed it makes it clear that what we were dealing with by this point was not so much evidence-based policy, but evidence-based *politics*, a different matter altogether. The idea that no compilation of facts, however well researched, can tell you whether you do or do not live in a good society, seems to have occurred to the report's authors only belatedly.

Louise Casey's somewhat more demotic remarks (made, to be fair, in the context of an after dinner speech) earned her much criticism, not so much because she challenged the shibboleth of EBP, but because as the then head of the Government's initiative to stamp out "antisocial behaviour," her response to being repeatedly told to "go and find the evidence," to back up any proposed policy initiative, was perhaps a little too robust.

Ms Casey's remarks reflected wider concerns about evidence-based policymaking, many of which this chapter will explore. But her primary concern appeared to be about the *difficulty* of providing evidence, rather than the *desirability* of it in all cases. A more informed debate not only about the role of evidence but also about its limitations is needed, and the case of creative industries is a very good place to start.

This chapter will briefly re-cap on the understanding of "creative industries" as a term and as an instrument of public policy. We will then look at how this policy area was adopted and implemented in regions and cities across the UK and how, in so doing, it became linked to wider questions, not only of culture, but also of social policy.

We will then discuss the drive for evidence-based policymaking and the concerns about the quality of evidence produced, in the cultural and creative industries, before the chapter goes on to consider the often uneasy relationship between evidence and ideology that has in fact characterised the last seven years.

Terminology in these areas is always fraught and this chapter is unlikely to escape unscathed by this tension. When talking about the creative industries, as defined below by the UK Department of Culture, Media and Sport (DCMS), I will, not unreasonably, use the term "creative industries." When talking about the wider cultural sectors, I will use the term, "cultural and creative" industries, or more simply "culture," which should be taken to *include* the creative industries.

In addition, the term "creativity," now tends to be accorded a special status in terms of the economy. I find this problematic, as I do not think we have a clear enough notion of what "creativity" is in economic terms and I do not think it is synonymous with innovation or novelty, as is sometimes suggested. I shall therefore try not to use it at all.

A brief history of the creative industries

The debate about creative industries in the UK was always accompanied by a stress on evidence—in this case two "mapping" documents released in 1998 and 2001 (DCMS 1998, 2001), carried out under the auspices of a specific government taskforce. These, for the first time, attempted to ascertain the size and structure of the cultural sectors in the UK, by this time re-branded as the "creative industries."

This re-branding, in contrast to many, was important, as it reflected a desire to stress the economic importance of these sectors—their contribution to wealth and to job creation—rather than their aesthetic or cultural qualities. It positioned them as drivers of economic success, not recipients of subsidy, and was tied into a broader strategy aimed at building a "competitive knowledge economy" (Department of Trade and Industry 1999) in the UK—still reeling in some parts from the effects of the rapid de-industrialisation of the 1980s.

The term has come to be widely used internationally (Cunningham 2003), but has never quite been loved at home. The obvious question it provokes is; aren't all industries creative? To which the answer is that some of them are, and indeed some activities within the creative industries are not very creative at all. The term is intended to be descriptive, but inevitably sounds evaluative.

Academics in general still prefer the term "cultural industries," but its association with a tradition of critique rendered it unsuitable for policymakers keen to stress the positive potential of these sectors. This tendency to boosterism has, as we shall see, infected the creative industries debate ever since and given it an ideological baggage of its own, associating it with much of the "new economy" hype of the dot-com (or dotcom) boom and bust (Miller 2004).

Further debate has also dogged the question of what is and isn't included in the creative industries. The original definition of a creative industry was woolly enough to allow for broad interpretations:

> activities which have their origin in individual creativity, skill and talent

> and which have the potential for wealth and job creation through generation and exploitation of intellectual property. (DCMS 1998)

Despite this stress on *individual* creativity as the source of wealth, the sectors that were included in the "mapping": advertising, architecture, arts and antiques markets, crafts, design, designer fashion, film, interactive leisure software, music, performing arts, publishing, software and computer services, television and radio, in fact ignore many individual creators, thus publishing is included but not writers, arts markets are included, but not artists.

Some felt that the definitions were too narrow (Howkins 2001) arguing that the sciences were clearly home to much creativity; others were concerned that they were too broad, the inclusion of "software" for example meant that everyone who worked for Microsoft in whatever capacity was included in the numbers; J K Rowling was not (Holden 2004).

Both debates have continued on and off since 1997 and while there is much to debate, both in definitions and in scope, the real achievement of the Mapping Documents and the Creative Industries Taskforce was to convince mainstream policymakers of the economic importance of these activities and thus the value of public investment in them.

The numbers that the Taskforce produced were, in truth, a surprise to many and have continued to be impressive. The creative industries grew by an average of eight per cent per annum between 1997 and 2001, more than twice the rate of the economy as a whole. By 2005, they employed more than two million people in the UK, accounting for a larger percentage of the UK's economy than that in any other country, according to government figures (Purnell 2005).

The inclusion of software as a category undoubtedly boosted the national numbers and there is some justifiable concern that in sectors such as music, film and design, UK performance has actually slipped in recent years (NESTA 2005). The lack of comparable international research on these questions makes it very difficult to assess relative national performance definitively and I have written elsewhere about the danger of over-hyping the economic contribution of the creative industries (Oakley 2004); but that is not to suggest that they are not important within the context of the UK or global economy.

The success of this economic case however led to greater scrutiny of the claims being made. Yes, the creative industries were creating more jobs in London than any other sector save financial services; but who is benefiting from those jobs? Ethnic minorities in particular remain under-represented in most sectors. Yes, we wanted to be a "creative hub" but in what sense was "Britain" benefiting when the vast majority of jobs were concentrated in London and the South East? The numbers were impressive, but as is the way with numbers, they raised a lot more questions than answers.

Creative industries in place—culture comes back

Despite the inclusion of performing arts and music under the rubric of creative industries, many worried that the development of this notion threatened traditional support for the arts in Britain and would lead to support only of those arts forms that could be shown to have a demonstrable economic impact (Ellis 2003).

Although those worries have continued, and tend to get quieter or louder in concert with the level of public funding for the Arts Council, in practice, something rather different happened.

Creative Industries, having been dreamt up by a central Taskforce reporting to the UK's culture ministry, went regional, and in doing so, the strong distinction that some would like to make between the subsidised cultural sectors and the market-facing creative industries (Heartfield 2005), was weakened.

Although the policy direction was centralised, the most enthusiastic backers of creative industries in the UK, and the channels through which most of the money flowed, were either in the nations, such as Scotland or Wales, or the English regions, such as the North West. This in part reflected the moves towards "devolution" in the UK, with some devolvement of political power and decision-making to the nations and much more limited systems for decision-making and policy formulation in the English regions.

Sometimes in concert with and sometimes in parallel to these regional moves, cities in the UK also showed renewed interest in the creative industries. Aspects of the creative industries, not least their close relationship with the urban milieu and their potential for spillovers into consumption and leisure industries, made them a particular focus for urban policymakers. It is not just that the creative sector is concentrated in the urban conurbations, but that much of its "support infrastructure" is inseparable from that of the urban milieu. . And in many cases, getting the right policy framework to support the urban milieu, whether on public transport or licensing laws, planning or migration, means getting the right policy framework to support a creative industrial sector (Leadbeater and Oakley 1999).

As creative industry policy and funding in the UK moved to regions and cities, therefore, the links to broader culture, not just the subsidised arts sectors, but also sport, tourism and the heritage sectors, which seemed to some to be missing from "creative industries," were re-instated in policy discussions and to some extent in joint funding, particularly in bodies such as "regional cultural consortia," established in each of the English regions. These bring together representatives from regional agencies dealing with subsidised arts, heritage, sport, tourism and libraries, as well as the creative industries.

In part this was practical. In most places in the UK, outside London, the South East and some of the larger urban economies like Manchester, the numbers of people employed in the creative industries are simply not big enough to warrant a huge amount of policy attention. In many areas of the UK, the public sector is the largest employer of graduates (Local Futures 2004) and in all parts of the UK the role of the public cultural sector, from museums and libraries, theatres and concert halls to publicly funded festivals and public art, is vital—not only for aesthetic or cultural reasons (though these are central), but also for the links that exist between the public cultural sector and the "commercial" creative economy.

Public spending on culture undoubtedly has wider economic impacts in the market, even if the mechanisms used to track such impacts are often flawed (Oakley 2004a). Consider just a couple of ways in which this takes place. Practitioners often move between the subsidised sectors and the market—British actors work in both the subsidised and commercial theatres, before stints on TV—again a mixed public and private market—or in some cases film, which in most countries receives partial public subsidy. Consumers, similarly, move between the commercial and the public sectors, often without knowing it, and the aesthetic sensibilities that are formed in public art galleries, in public education and in urban design, often exercise themselves in the market through the increased consumption of cultural goods and services.

Even those sectors where public funds appear to have very little role—popular music or video games for example—still owe much of their vibrancy to public policy. The combination of policies on education, urban design, licensing laws, migration and drugs that produced the "Mersey Sound" or "Madchester" is dense and difficult to unpick—but that does not make it irrelevant. In truth, a creative industries sector that had no links to public culture would likely consist of advertising, bits of the software market, and possibly, crafts—hardly the stuff of a global "creative hub."

But in re-instating the links between the creative industries and "culture" would also see the re-introduction into creative industries policy of many of the problematic issues it thought it had left behind in cultural policy.

As Gibson and Klocker have argued in the context of Australia, the notion of "creativity" was adopted by governments pursuing regional economic development, in part because:

> of its ability to act as catalyst in the cultural transition of individuals from "citizens" into "entrepreneurs" and "consumers," the idealised companions of the neoliberal state. (2005, 33)

This was equally true in the UK, as was the notion that economic development required "competition" between places and that, when it came to creativity, "all places must play against each other." But in the case of New Labour and of the UK at the turn of the millennium, an unquestioned focus on market-based competition in economics was forced to sit alongside a social policy which articulated "inclusion" as its goal.

Instead of being able to concentrate on the purely economic—jobs and growth—questions were being asked about social impacts, concerns were raised about gentrification, and, in cultural terms, the usual accusations were made about "dumbing down." All this, and policymakers kept being asked for more and more "evidence"—the production of which, as we shall see, is anything but trivial.

"Just the facts, m'am"—the hunt for evidence

Developing evidence across sectors as diverse as the creative industries, and in particular developing evidence of the increasingly wide range of impacts that politicians were now seeking, was always going to be problematic. The UK's culture department (DCMS 2003) acknowledged that compared to healthcare, evidence across all other policy areas was seriously underdeveloped. However, the Ministry were not persuaded that what they described as the "cultural and creativity," areas faced unique barriers in the construction of an improved evidence base.

The problems with statistical data on the creative and cultural sectors are relatively well known and not that different from those in other areas of the services economy. In particular, the inability of the Standard Industrial Classifications (SIC) system to keep up with changes in the economy means that tracking down the numbers of cultural workers is difficult, there being no categories at all for occupations such as video games designer or DJ. In the UK, many of the more intractable problems refer to data at sub-regional and local levels, where sample sizes get very small—though this is often the level at which policy interventions take place. In particular, in the creative and cultural industry sectors, where part-time and self-employment is widespread, data on self-employment is often considered unreliable, with many countries, including the UK, only keeping a record of those who pay tax, and hence earn above a particular income threshold. As incomes in these sectors vary widely, but include many on relatively low incomes, such people do not often make it into the count.

Most researchers working in this area have concluded that in the short to medium term the unreliability of data at the local level means that official statistics need to be supplemented by trade associations' data and commercial market research, though the latter is often very expensive for public policymakers to collect.

There also is a range of difficulties in assessing *indirect* economic benefits. In particular, the use of multipliers has drawn much criticism of economic impact studies. There is often simply confusion over how to define a multiplier (Johnson and Thomas 2001) and practice varies widely. Second, there is the problem of opportunity costs—if a festival generates spending in the local economy is that more or less spending than would be generated by another type of event? And finally there is the problem of long-term studies—in practice multiplier effects are often used in bid documents to estimate economic benefits, but as Adrian Ellis points out:

> estimates used for funding proposals are almost never compared against actual outcomes, so that some empirical basis can be built up to test assertions. (2003, 11)

When it comes to qualitative evidence, the issues are, if anything, even more fraught. The use of "anecdotal" evidence in policymaking remains contentious, as Ray Pawson comments:

> whilst it is hardly obsessed with the lofty ambition of qualifying for the inner sanctum of "science," the very idea of evidence-based policy rests on the matter of differentiating its efforts from "common sense," "intuition," experience," "value judgements," and so on. (2003, 4)

Others argue that in always valuing "hard data" above the "anecdotal" we simply privilege certain kinds of knowledge: the scientific and the rational, which are usually the preserve of cultural and economic elites, above the more informal or intuitive "local knowledge." Or that in trying to impose a quantitative measure on everything, we may end up concluding that something (such as participation in an arts program) has no effect at all, simply because we are unable to measure it.

The official UK government view (Davies 2004) is that single studies, case studies and public opinion surveys do indeed have a role as evidence, if carried out to "the highest possible standard." The issue then becomes one of standards, not just methods, of evidence-gathering, though it seems clear that much of the research carried out on the creative sectors falls short of the required standards (MLA 2005). This is particularly true of "case studies," a term which is frequently used simply to describe an interview with a single individual. The subjects for interview are rarely chosen as systematically as even the most basic quality standards would require.

As Coalter argues, much of the case study data is based on descriptions of output, with outcomes often being assumed (2001, 7). Research is often too short term, sometimes project-specific, and the requisite follow-up work to determine longer-term outcomes is never carried out. These shortcomings do not lead to a diminution of claims however, and a small, short-term impact on a sub-group of people is often presented as though it were an enduring impact on a much larger group.

As in economic impact studies, there is rarely an attempt to measure opportunity costs: the benefits of spending money on one particular intervention, rather than others. The real question for policymakers is often not, "did this work?" but "did this work better than another approach?." This is not to minimise the difficulties in answering such questions. Detailed social experiments of this type— comparing one set of activities with another—are expensive and can be difficult to construct, but an understanding of the issue of opportunity costs should be part of any serious impact research.

When looking at place-based interventions, and thus at effects on "regeneration," an even more deep-seated problem is one that is common across social science research, known as the "problem of aggregation." This refers to the difficulty inherent in linking micro-level effects on individuals to the macro level of the community. Guetzkow (2002) makes it clear that the mechanisms by which individual effects are translated into community-level effects, or even how researchers hypothesise they may be translated, are problematic and contested.

Several approaches to this have been tried including: simply adding up the percentage of individuals in a population that are affected by something (more being better); looking for the threshold or "tipping point" whereby individual effects become community effects; looking at the type of networks that people form via cultural activities (ie. do they increase their social capital?), or looking at how a few key individuals or community leaders are affected and how this feeds into a general climate of opinion. Much published impact research, however, simply assumes that an individual impact can become a group impact, without any explanation of the mechanism involved.

In many of these instances then, the primary problem is the gulf between the *claims made for a particular activity* and *the evidence to substantiate those claims.* This can be summed up as "the advocacy problem": a widespread critique that research in this area is rarely impartial and is too often confused with advocacy (Selwood 2002).

The advocacy problem is often seen as one that belongs to artists and those campaigning for more public spending—either directly on the arts or on physical regeneration and new initiatives. But another explanation of what is driving the confusion between advocacy and research in these areas is that public policy is itself confused. This is not to say that public policy in an area cannot have multiple aims—an investment in an arts centre for example could have social, economic and aesthetic aims all at once and there would be no necessary contradiction there. The confusion in current policy seems more profound.

New Labour, having set itself up as being explicitly non-ideological—claming that "what matters is what works," is often surprisingly deaf to the evidence that treasured areas of social or economic policy are not working, or not working in the way they were meant to. In the case of the creative industries the confusion of social and economic policies means that very fancy footwork is required if the "right" answers are to be produced.

At home with the creative class

It is impossible within a piece of this length to do justice to the amount of recent research on the social and economic impacts of the creative and cultural industries. While the concerns above remain valid, the evidence base is improving as it grows globally (Oakley 2004b). But given the argument that places, and in particular cities, have become central to the discourse of creative industries, it is worth just tracing some of the evidence that recent research in this area has produced. In particular, how does it tie into the Government's desire to see "culture at the heart of regeneration?" (Evans and Shaw 2004, 2)

For many people the notion of culturally-led regeneration is firmly tied up with the idea of iconic or destination buildings and as Baniotopoulou points out, many of these buildings are art, particularly modern art, museums (2001,7). The Guggenheim in Bilbao is perhaps the best example of this in Europe, and it is one that many other cities have tried to emulate in the past decade. (Hannigan 2004; Evans 2003)

As Beatriz Garcia (2004) points out, the legacy of the Barcelona Olympics and the Seville Expo meant that Spain in the early 1990s saw urban regeneration very much as an outcome of flagship, high profile capital projects. The Guggenheim—a cultural icon in a depressed, post-industrial city—fitted this picture exactly and was joined by other major infrastructure projects, notably a high quality underground rail system designed by Norman Foster. As Garcia suggests, the short-term benefits were impressive: a worldwide transformation of the city's image, indicated by a large rise in tourism. Overseas tourism was up 42.7 per cent between 1994 and 2000 and Spanish tourism up even more, by 58 per cent. (2003)

But in recent years questions have been asked about the sustainability of such high-profile initiatives. In particular, does the tourism impact last beyond first time visits? Does it translate into other local jobs? And in the case of an overseas investor like Guggenheim—does it provide a platform for local artists or simply act as a franchise of the US parent institution?

Baniotopoulou (2001) is very clear on this last point, arguing that the presence of the Guggenheim in Bilbao has done little for the local arts scene, featuring relatively few Basque or Spanish artists among its acquisitions, and reflecting instead the desire of the Guggenheim Foundation itself to be at the forefront of the international art world.

Evans (2003) argues that this lack of a link between "flagship" cultural projects and local creative businesses is not confined to Bilbao, but is a weakness of *grands projets culturels* everywhere. Indeed he goes on to argue, along with others (Baniotopoulou 2001), that such flagship projects are often undertaken *at the expense of* local and regional cultural development.

In addition, as David Harvey argued over a decade ago, despite their relatively high costs, such flagship projects can in fact be relatively easily reproduced in different locations, "thus rendering any competitive advantage within a system of cities, ephemeral." (1989, 44)

The challenge of sustaining such high profile initiatives (and the natural limit to the number of iconic buildings) means that the policy focus is instead shifting towards supporting smaller, more localised regeneration projects, which seek to combine both production and consumption. (Hannigan 2004)

Mommaas describes this as a "more finely tuned" policy of "cultural clustering," that is "aimed at creating spaces, quarter and milieus for cultural production and creativity." (2004, 11) As Evans and Shaw (2004) point out, these aim to take advantage of information and knowledge sharing, as well as what geographer Michael Storper calls, "local buzz." (Venables and Storper 2003, 22) Local buzz strategies are concerned with increasing the density of relationships between firms *within* a geographic area and providing specialised support and services, ranging from business support to consumption activities for them *in situ.*

Although many of these strategies are focused on small creative businesses—amenities such as galleries (public and private) are often used to complement these production centres, as has been the case in several UK cities and others from Brisbane to Toronto and Hong Kong. (Evans and Shaw 2004; Shorthose 2004)

While much of the research on cultural regeneration has focused on the role of artists and cultural entrepreneurs, one of the most influential of recent writers on this topic has instead looked at the ability of cultural investments to attract a much broader class of "knowledge worker."

The publication in 2002 of Richard Florida's *Rise of the Creative Class*, gave a temporary fillip to urban policymakers who had been looking for ways to make the "case for culture," though many commentators, myself included, would come to feel it was an unwelcome one.

Florida's book and his other work over more than a decade, builds on the insights of urbanists (Jacobs 1992; Hall 1998) and "new growth theorists" (Romer 1994) about the importance of human capital in a knowledge-based economy and the roles of cities and city-regions in attracting that capital.

First, he argues, cities provide "thick" labour markets—a depth of skills and talent, together with the associated institutional infrastructure (education, business support, networks) —for firms to draw on. This leads to greater innovation than a firm on its own would be capable of, as innovation involves many people engaging in dynamic, collaborative processes, face-to-face interaction and the transmission of codified and tacit knowledge. (Porter 2000)

Second, cities have well-developed and attractive urban amenities, though in Florida's case the valued ones seem to be mostly suited to the aesthetic sensibilities of younger professionals, rather than a broader class of citizen. (Bradford 2004.) Third, urban culture is defined by what he calls its "tolerance of diversity" (he particularly cities tolerance of gay lifestyles here) and by the vibrancy of its "street scene" reflected in cafes, clubs, bars and galleries.

Florida's contribution was to link this to the mobility of his "creative class," a broad group ranging from those employed in science and the arts to financial services and health care (much broader than those who work in the creative industries), to argue that in a consumer economy this class will pursue employment opportunities less on the qualities of the employer itself, and more on the amenities and lifestyle options of the city or region where potential employers are located.

His other distinctive contribution is his attempt to quantitatively measure, through the uses of "indexes" of various sorts, the conditions that he deems necessary for successful city-regions. Both the "gay index" and the "bohemian index" are said to measure a region's tolerance and lifestyle diversity. His "gay index" measure is of coupled, same sex households living in a particular metropolitan area. The numbers are drawn from the US Census and have been criticised for including same-sex households that are not gay. The UK Census does not contain a question about sexual orientation, so UK researchers attempting to do the same thing, as in Demos' 2003 "Boho" Index, are forced to use proxies (number of gay clubs, bars, networks, support groups, businesses etc), which are obviously unsatisfactory.

Likewise his "bohemian index" attempts to measure the number of individuals employed in "artistic and creative occupations," in a particular area. But as discussed earlier, data about creative occupations in the UK is very unreliable at local level.

Like all researchers into human capital—Florida is faced with having to use qualifications as a proxy for skills in his talent index, thus not capturing the talents or skills of those without formal education qualifications. However, given that the aim is to compare region to region, this will allow for a comparative, if not numerically accurate, figure in each case.

Florida's work has been undeniably influential, but there are those that worry that its rapid adoption into policy outside the US reveals a lack of recognition of the different political, social and economic circumstances that prevail elsewhere. (Gertler 2004; Nathan 2005)

First, and perhaps of most practical concern for regions in the UK seeking to use this formula, is the fact that Britons, in common with other Europeans, are much less geographically mobile than people in the US. Only about ten per cent of households move every year in England, of whom only about one per cent move between regions. (Donovan et al. 2002) Thus trying to regenerate, for example, the North East of England (which has suffered serious industrial decline), *solely* via a strategy of attracting the "mobile creative class," not only risks problems associated with gentrification, but is unlikely to succeed in attracting people in sufficient numbers.

As Douglas and Morrow (2003) point out, many of the practical interventions that have resulted from Florida's work have focused on simply marketing the consumption value of "bohemian" districts; concentrating on short term physical improvements to particular neighbourhoods, while at the same time under-investing in the essential amenities (from good schools to good public transport) that help sustain city-regions. The argument is that simply focusing attention on "the creative class" (for which read young, single men) may lead to neglect of other policies, from affordable childcare to good schools, that may apply to older or female workers. (Bradford 2004)

In addition, there is a tendency in Florida's work to conflate tolerance with cultural consumption. An interest in ethnic food or world music is not the same as genuine racial or religious tolerance and other indicators need to be taken into account, such as racial divisions in the labour force, or number of attacks on racial or religious minorities, before declaring a place as "tolerant." His argument that creative class workers are attracted to places that they perceive as tolerant suggests a similarity of politics and values among these workers that is directly contradicted by other research on this subject. (Nathan 2005)

As Neil Bradford argues, "in celebrating cultural diversity, Florida pays much less attention to the reality of racially-divided urban labour markets and the fact that some of his creative hot spots are also socially polarised places." (2004, 4) When that does happen, a less optimistic picture of Florida's creative economy appears, as he himself admits in later writings. (Florida 2004) His colleague Kevin Stolarick has developed an index of wage inequality which compares the wages of the creative class to others and reveals that, in the US at least, city-regions that rank highest in terms of creative economic strength also rank highest in economic inequality.

Since the 1990s, this link between the growth of knowledge-based economies and increasing economic inequality was becoming apparent across developed economies. (Atkinson 2005) As Graeme Evans has commented about the BritArt-led boom in Hoxton in the East End of London,

> 1,000 local jobs a year have been created, but the local unemployment level never seems to change. Partly thanks to the success of Hoxton, land values in the area have soared. So locals who do get jobs often have to move outside the borough. (2004, 90.)

In response to this, some local authorities and other public agencies have sought to take a more pro-active stance to ensure that culturally-led economic development is more balanced and less in danger of driving out the creative entrepreneurs (and other locals) on which it depends. (Gertler 2004; Bottomley et al. 2003*)*

Cameron and Coaffee in their 2004 study of Gateshead argue that there is a distinction between cities where gentrification is driven by commercial capital (such as the Hoxton example) and those where what they call "positive gentrification," is driven by public authorities. The latter, they argue, is more relevant in cities affected by de-industrialisation such as the North East of England, where:

> private capital has to be dragged kicking and screaming into de-valourised urban locations through the initiative and investment of the public sector. (2004, 23)

Meric Gertler argues that if we reject the notion of creatively-led urban regeneration as simply the "plug-and-play-ground for twenty and thirty-something members of the creative class," then we need to adapt this thesis to our specific political and social context and, in particular, to the challenge of developing more economically and socially sustainable communities. (2004, 30) The desire to build "socially inclusive creative places," Gertler argues, has to be a conscious choice on the part of policymakers—inclusivity is not inherent in culturally-led regeneration. He could have added that it needs to be a political choice, one that cannot be arrived at by a process of collecting evidence.

Conflict and contradiction—the limits of evidence

As demands for better and clearer evidence of social and economic impacts has grown in the UK, so too have criticisms of the Government's whole approach to evidence-based policymaking. Some argued that social and economic policy is simply too complex to be subjected to processes designed for the laboratory (Marston and Watts 2003) and that notions of "proof" and "evidence" had simply been lifted from the sciences and transposed to public policy without the necessary adjustments. In the case of culture, in particular, the instrumentalism of looking at impacts beyond the aesthetic is seen by many as suspect. (Ellis 2003) Others are concerned that the available data does not support a rigorous process of research. (Selwood 2002) Many others are also sceptical of the inconsistency of an Administration that routinely ignores evidence in criminal justice, drugs policy and intelligence reporting, to name but three areas, while demanding better evidence from cultural and creative practitioners.

In the case of urban policy, an even more valid critique may be that the Government has continued to pursue culturally-led regeneration *despite* evidence not only that it might it not work, but that it might deliver precisely the opposite from what was intended—namely cities that are even more polarised by race, class and education than is currently the case.

In her 2001 review of the literature on urban development policies, Betsy Donald revealed two contradictory themes. The first was the argument that economic growth was antithetical to quality of life and sustainability. The second, along the Florida lines, was that quality of life was vital to economic growth, particularly of cities. This is reflected strongly in the current policy rhetoric on cultural regeneration—culture must both generate growth and save us from the problems of growth, whether that is soulless suburbs built to house new knowledge workers or stress caused by a long working hours culture.

This is not to deny that the cultural and creative sectors may indeed play both these roles, as commentators such as Gertler (2004) and Polese and Stren (2000) argue that they can, but simply to observe that policy as currently articulated in these sectors suggests that there *is no tension or contradiction between these roles*, any more than there is tension between industries which sell themselves in part on the basis of "diversity," while containing relatively few workers of ethnic minority—at least in any powerful positions.

New Labour, creativity and the poor

The stress on evidence is a strong feature of "Third Way," politics, a sometimes incoherent set of ideas that briefly united centre-left parties such as the UK Labour and Australian Labor parties, the US Democrats, and, at least until the Iraq war, some elements of the German Social Democratic party, among others. (Giddens 1998)

Australia's one time Labor leader, Mark Latham in a discussion on welfare reform clearly articulated the role that evidence had come to play for Third Wayers:

> The myths of the welfare states are based on old ideological ways of thinking, struggle between government first and market first policies. It is now clear that both approaches are flawed. The world has moved on. Welfare policymakers need to look beyond the old left and the new right to those evidence-based policies that can end the human tragedy of poverty. (Latham 2001, 42)

This expresses many of the cherished beliefs of the Third Way: that ideology is somehow bad in politics, that previous political approaches have "failed," that progress and modernity must be defended, that evidence and facts are neutral and speak for themselves, and that only through their application can we solve particular social problems.

What is lost in this approach is the political theory underlying it, without which judgements such as the "tragedy" of poverty or the "failure" of state and markets are meaningless. It is not evidence that makes us think that poverty is bad; it is an *a priori* political or moral belief that it is bad. We may then look at evidence to help us find ways to tackle poverty, but we should not approach it the other way round. As Brian Barry says, "No compilation of "facts" can tell us about the fairness or unfairness of a tax system. For that we have to have a theory of social justice." (2005, 13)

A closer look, therefore, at Latham's text suggests that it expresses very clearly an ideology, and one that has guided UK policy on culture and the creative industries, among other things.

This could perhaps be simply expressed as the notion that previous governments of the left had neglected economic growth, or at least treated it as incidental to their basic concern with economic security and redistribution. (Diamond and Giddens 2005) Third Way governments were much more "pro market"; indeed Labour in the UK had followed many Conservative economic policies in areas like public spending and privatisation of public services. But Third Wayers retained a residual concern with poverty, uncomfortable at some level with the evidence that neoliberal economic policies would lead to more of it. The core tenet of neoliberalism—the supremacy of the market—could not be challenged; ways must therefore be found to obfuscate this contradiction. The issue here is not why economic policy encompassed social policy in this way, but why the discourse of "culture and creativity" was at the front line of this?

Conclusions

The value of the "creative industries" as a distinct set of economic activities, promising high growth rates, job creation, and links to a wider notion of the "knowledge-based economy," has taken root in the UK and globally. It has successfully embodied the notion that the arts, media and design, particularly via their links to the urban economy, are drivers of economic growth, not just recipients of public subsidy.

These sectors have attracted a good deal of attention from regional and urban policymakers in the UK and elsewhere, keen to reverse the economic damage of de-industrialisation and to use local, cultural assets to develop a more competitive economy. Thus an economic case for the creative industries—that they promote jobs and GDP growth—has inevitably become broader, tied up with the creation of sustainable urban milieu. And the links between the creative industries and the wider cultural assets of a place have been restored in policy terms, though not always explicitly.

The UK's New Labour government in particular has supported the idea of the creative and cultural sectors as both a source of economic growth and a balm for the casualties of the economy—whether this is declining inner cities areas or marginalised communities. In many ways, creative and cultural industries have been the ideal policy area for a "Third Way" politics—promising "win wins" on both growth and social justice.

But advancing these arguments means turning one's back on another New Labour shibboleth—that of the primacy of evidence. Much of the evidence collected so far suggests that the development of creative industries, while it may offer many benefits for regional economies, is unlikely to promote social inclusion and greater equality.

As Gibson and Klocker have argued, "creative economy discourse is contradictory, becoming more difficult to sustain precisely as it seeks hegemonic status," (2005, 98) citing tensions such as those between competition and co-operation, socialisation and the neoliberal primacy of the individual consumer, and "diversity" versus the gentrifying impacts of these developments. Nowhere are these contradictions more evident than in New Labour rhetoric about the creative industries.

References

Atkinson, Robert. 2005. Inequality in the new knowledge economy. In *The New Egalitarianism*, ed. Anthony Giddens and Patrick Diamond. Cambridge: Polity Press.

Baniotopoulou, Evdoxia. 2001. Art for Whose Sake? Modern Arts Museums and their Role in Transforming Societies: The case of the Guggenheim Bilbao. *Journal of Conservation and Museum Studies* 7: 1-15.

Barry, Brian. 2005. *Why Social Justice Matters*. Cambridge: Polity Press.

Bradford, Neil. 2004. Creative Cities. Background Paper F/46. Canadian Policy Research Networks.

Bottomley, Alison, Monica MacDonald, Karen Poetker, and Etolie Stewart. 2003. 401 Richmond.
http:// www.401richmond.net/401_Richmond.pdf.

Cameron, Stuart, and Jon Coaffee. 2005. Arts, Gentrification and Regeneration—from artist as pioneer to public arts. *International Journal of Housing Policy,* 5: 41-60.

Allison, Mary, and Fred Coalter. 2001. *Realising the potential of cultural services, the case for museums.* Local Government Association.

Cunningham, Stuart. 2003. *The Evolving Creative Industries.* Queensland University of Technology.

Davies, Phillip. 2004. *Is Evidence-Based Government Possible*? Jerry Lee Lecture 2004.

Department for Culture Media and Sport. 2003. *A Research Strategy for DCMS*, Technical Paper No 3, June 2003, London.

Demos. 2003. *Boho Britain Creativity Index.*

Department of Trade and Industry. 1998. Our *Competitive Future: building the knowledge driven economy.*

Donald, Betsy. 2001. Economic Competitiveness and Quality of Life in City Regions; compatible concepts? *Canadian Journal of Urban and Regional Research,* 10, 2: 259-274.

Donald, Betsy, and Douglas Morrow. 2003. *Competing for Talent: Implications for Social and Cultural Policy in Canadian City-Regions.* SRA, Department of Canadian Heritage.

Donovan, Nick, Tony Pilch, and Tracey Rubenstein. 2002. *Geographic Mobility.* PIU Discussion Paper.
http:// ww.pm.gov.uk/output/page699.asp.

Ellis, Adrian. 2003. Valuing Culture. Paper presented at Demos seminar, Valuing Culture, 17 June 2003 in London.

Evans, Graeme and Phyllida Shaw. 2004. *The Contribution of Culture to Regeneration in the UK: A Review of the Evidence.* http://culture.gov.uk/.

Evans, Graeme. 2004. Cultural Industry quarters: from pre-industrial to post-industrial production. In *City of Quarters: Urban Villages in the Contemporary City*, ed. David Bell and Mark Jayne. Aldershot: Ashgate.

Florida, Richard. 2002. *The Rise of the Creative Class*. New York: Basic Books.

—. 2005. Creative Class War, How the GOP's anti-elitism could ruin America's Economy. *Washington Monthly,* Jan/Feb.

García, Beatriz. 2004. Urban Regeneration, Arts Programming and Major events: Glasgow 1990, Sydney 2000 and Barcelona 2004. In L Gibson and D Stevenson, eds., *Special Issue of the International Journal of Cultural Policy: Urban Space and the Uses of Culture* 10, 1: 103-118.

Gertler, Meric. 2004. *Creative Cities: What are they for, how do they work and how do we build them?* Canadian Policy Research Networks, Background Paper F/48.

Gibson, Chris and Natascha Klocker. 2005. The "Cultural Turn" in Australian Regional Economic Development Discourse: Neoliberalising Creativity? *Geographical Research*, 43, 1: 93-102.

Giddens, Anthony. 1998. *The Third Way, the Renewal of Social Democracy*. Cambridge: Polity Press

Giddens, Anthony and Patrick Diamond. 2005. *The New Egalitarianism.* Cambridge: Polity Press.

Guetzkow, Joshua. 2002. *How the Arts Impact Communities: An introduction to the literature on arts impact studies.* Centre for Arts and Cultural Policy Studies, Working Paper Series 20. Princeton University.

Hall, Peter. 1998. *Cities in Civilisation*. London: Weidenfeld & Nicholson.

Hannigan, John. 2004. Boom Towns and Cool Cities: the perils and prospects of developing a distinctive urban brand in the global economy. Paper presented at the Leverhulme Symposium on The Resurgent City, April 20. London: London School of Economics.

Harvey, David. 1989. From Managerialism to entrepreneurialism: the transformation in urban governance in late capitalism. Reproduced in *Spaces of Capital*, 2001. Edinburgh: Edinburgh University Press.

Heartfield, James. 2005. The Creativity Gap. *Blueprint Magazine*, May.

Hui, Demond, NG Chan-Hung and Partick Mok. 2004. A Study on Hong

Kong, Creativity Index, Interim Report. Hong Kong Home Affairs Bureau.

Holden, John. 2004. *Capturing Cultural Value: How culture has become a tool of government policy*. London: Demos.

Howkins, John. 2001. *The Creative Economy: How people make money from ideas*. London: Penguin.

Jacobs, Jane. 1992. *The Death and Life of Great American Cities*. New York: Vintage Books.

Johnson, Peter, and Barry Thomas. 2000. *Assessing the Economic Impacts of the Arts*. University of Durham (unpublished).

Kotkin, Joel. 2005. Urban Legends: Cities aren't doing as well as you think. *New Republic,* May 23.

Latham, Mark. 2001. Myths of the Welfare State. *Policy*, 17, 3: 40-43.

Leadbeater, Charles, and Kate Oakley. 1999. *The Independents: Britain's new cultural entrepreneurs*. London: Demos.

Local Futures. 2004. *Accidental Keynesianism: The public sector in the knowledge economy*. Available from Local Futures Group, http://www.localfutures.com/.

London Development Agency. 2004. *Creative London*.

The Museums, Libraries and Archives Council (MLA). 2005. *New Directions in Social Policy, Developing the evidence base for museums, libraries and archives.*

Marston, Greg, and Rob Watts. 2003. Tampering with the Evidence, a critical appraisal of evidence-based policymaking. *The Drawing Board, an Australian Review of Public Affairs*, 3, 3: 144-163.

Miller, Toby. 2004. A View from a Fossil: the new economy, creativity and consumption—two or three things I don't believe in. *International Journal of Cultural Studies,* 7, 1: 67-77.

Mommaas, Hans. 2004. Cultural Clusters and the Post-Industrial City: Towards the remapping of urban cultural policy. *Urban Studies,* 41, 3: 507-32.

Nathan, Max. 2005. *The Wrong Stuff, Creative Class Theory, Diversity and City Performance.* IPPR Centre for Cities. Discussion Paper No 1.

National Endowment for Science, Technology and The Arts. 2005. *Creating Value, how the UK can invest in new creative businesses.*

Oakley, Kate. 2000. *E-government, A comparative Study of Online Government*. Published by Cable & Wireless. Available from author.

—. 2004a. Not so cool Britannia, The role of the creative industries in economic development. *International Journal of Cultural Studies*, 7,1: 67-77.

—. 2004b. *Developing the evidence base for support of cultural and creative activities in South East England.* Available from http://www.semlac.org.uk/docs/Oakley%20Report%2012.4.04.pdf.

Pawson, Ray. 2003. *Assessing the quality of evidence in evidence-based policy: why, how and when?*. The Economic and Social Research Council, Research Methods Working Paper No 1.

Polese, Mario, and Richard Stren. 2000. *The Social Sustainability of Cities*. Toronto: University of Toronto Press

Porter, Michael. 2000. Locations, Clusters and Company Strategy. In *The Oxford Handbook of Economic, Geography* ed. Gordon Clark, Maryann Feldman and Meric Gertler. Oxford: Oxford University Press.

Purnell, James. 2005. Making Britain the World's Creative Hub. Speech to the Institute for Public Policy Research, June 16.

Romer, Paul. 1994. The Origins of Endogenous Growth. *Journal of Economic Perspectives*, 8, 1: 3-22.

Selwood, Sara. 2002. Measuring Culture. Available at http://www.spikedonline.com/Articles/00000006DBAF.htm.

Siegel, Fred. 2005. Floridizing the World. *Weekly Standard*, February 28.

Venables, Anthony, and Michael Storper. 2003. Buzz: Face to Face Contact and the Urban Economy. Paper presented at the Leverhulme Symposium on The Resurgent City, April 29. London: London School of Economics.

Zukin, Sharon. 1982. *Loft Living, Culture and Capital in Urban Change.* New Jersey: Rutgers University Press.

CHAPTER THREE

CREATIVE ARTS, PEOPLE AND PLACES: WHICH POLICY DIRECTIONS?

CHRIS GIBSON

Introduction

Arts communities face an ever-more peculiar scenario: money is tight (as always) and funding support for arts initiatives is hyper-competitive, while job security for arts workers remains largely insecure. But at the same time, there is a new kind of demand for the arts. The rise of a more commercially-orientated "creative places" policy direction has given arts administrators and institutions new arguments to help make a case for their significance. The idea that the arts are vital to local, regional and urban economies is one capable of being appropriated by a diversity of actors (from museums and galleries to youth employment agencies), when seeking funding support and sympathy from funding bodies. The embrace of the arts "as creative economy" has generated new policy directions (O'Regan 2001; Gibson L 2001), and to some extent has brought about a blurring of boundaries between the responsibilities of regional development promoters and arts policy makers—whose interests were once generally considered quite separate.

This convergence (an "economic turn" in arts policy commensurate with a "creative turn" in regional economic development policy) has been fuelled by the popularity of key books—most notably Charles Landry's *The Creative City: A Toolkit for Urban Innovators* (2001), and Richard Florida's *The Rise of the Creative Class* (2002)—that have become "tomes of wisdom" of sorts among both economic development planners and arts policy makers. In the former, Landry provides a range of case studies of ways in which planning has been reinvented by creativity; in the latter, Florida argues that a key demographic group ("the creative class"—who include workers in the arts, design, multimedia and "content" industries) increasingly determine the future of places through their migration patterns, investment decisions and entrepreneurship. The influence of such books, and resulting ideas of the arts and creativity as economic stimuli, have filtered through to the Australian setting, where several local councils, as well as state government departments, have sought to develop cultural policy initiatives aimed at exploiting the arts for economic gain. In particular, the arts and creative industries are becoming part of strategies for economic recovery, for improved social capital and employment generation for young people.

My aim in writing this chapter is to provide some critical discussion of these policy directions. It is not to disempower those hoping to use ideas of arts-led recovery in important and progressive ways—I would for instance fully appreciate their strategic adoption where such arguments are needed to support fledging or vulnerable arts activities. Instead, I hope to generate debate around a number of practical and philosophical problems that I see spawned by the wholehearted and uncritical embrace of the "creative places" policy agenda (at least, in the manner in which it seems to be unfolding in some locations). This article outlines my take on some of these problems, and what a genuinely inclusive and equitable creative arts policy might require.

The arts: cultural *and* economic policy for places?

Since Paul Keating's *Creative Nation* policy statement was released in 1994, governments at national, state and local scales have (slowly, but surely) come to re-interpret the arts in terms of their commercial benefits to places. In many respects this is a facet of the increasingly neoliberal attitude of governments: the arts, according to neoliberal thinkers, are just like any other "market"; they ought to be driven by market forces, and thus the arts and arts institutions are judged by their capacity to grow and survive in the commercial world. Linked to this kind of thinking is also the realisation that the arts can indeed be "big business" (not least in places like Bilbao and Paris where associated tourist dollars are significant).

More subtly, researchers have recently argued that the arts benefit the economies of places in more indirect, but no less substantial, ways. Arts institutions are hubs for all kinds of workers (designers, architects, producers, directors, writers) and activities that add to the reputations for places; that make them more liveable, vibrant and interesting. Such reputations in turn translate into place popularity, in-migration of new residents and ideas and thus, arts scenes can be of enormous benefit to regional development strategies. The exemplar of this kind of argument is Richard Florida's "creative places" hypothesis. This is that certain places, those known as "creative", out-compete those that are not considered creative—that are considered drab, dying, old-fashioned or monotonous. Linked to this, a particular demographic group—the creative class—are identified as a social group that most contributes to the designation of places as "creative". The creative class is a social group made up of those directly innovating in their daily working lives, often producing cultural content (artists, musicians, designers, architects); as well as those that are avid consumers of creative products, including members of subcultures, young affluent professionals, and people of non-heterosexual preference. The creative class are drivers of employment growth—key to the diversification of economic base—and underpin the future in-migration of comparatively wealthy residents who bring niche industries, talent and entrepreneurial leadership (Florida 2002).

Arguments about the significance of the arts are increasingly being nestled within this narrative. Artists are the core of the creative workforce (Throsby 2001), and provision of the arts is highly desired by the creative class (via museums, galleries, live venues and so on). The creative class, it is said, consider the local presence of an arts scene when making their choices about residential location and investment. The arts also generate certain forms of employment and can provide a basis from which urban regeneration (culminating in main street café culture) can take place. In these ways, arguments about the arts have taken a decidedly commercial bent.

In Australia, governments at national and state scales have encouraged places to be more competitive with each other, and the arts have become a means for this. Where places have struggled to compete (for whatever reasons), governments have often put forward the arts as an alternative means of development. The arts have a certain amount of glamour, offer prospects of more interesting and diverse jobs, and have obvious links to tourism, niche retailing and other multiplier benefits. Success in the arts—and thus in the creative economy—has become a goal of strategies that seek to reinvigorate confidence in the capacity of locations to compete for mobile capital and new residents, to perform well in emerging industries, without the need for government subsidies or strong spatial and fiscal equalisation.

This conjunction of arts/cultural planning and economic development policy-making is becoming most common, though, at the local government level. Cultural policy has not been a central concern of local government in the past (hence the age-old "rates, roads and garbage" mantra on the role of local councils). More recently, local government has taken on responsibility for developing and implementing cultural policy initiatives, in part a reflection of the perception that local government is closer to the community, and thus can better respond to the specific cultural needs of its citizenry than other levels of government. In New South Wales, local governments have also been strongly encouraged to undertake cultural planning that includes strategies for the arts and creative industries (NSW Department of Local Government 2004). In 2001 that state's Ministry for the Arts established the "Western Sydney Local Government Arts Incentive Fund" which required the 14 local governments in Greater Western Sydney to undertake cultural planning as the precondition for funding. (NSW Ministry for the Arts 1999)

This shift towards arts and cultural planning at the local scale creates a situation in which key decisions are made about interpretation and implementation of policy directions. For some local councils, cultural planning has been absorbed under existing banners of social services or community cultural development. It has formed another means to embrace cultural diversity (and thus better deliver basic services through tailored information and consultation projects), and initiate special events, festivals, sponsorship of local arts projects, place-making schemes—including public art strategies—and subsidies for creative activities through council-owned performance spaces, studios and galleries (Gibson and Homan 2004). For other local councils, where a reorientation in governance towards entrepreneurial priorities has gained momentum in recent years, more overtly commercial discourses are brought to bear when developing cultural plans, reflecting what Miller observed as a "penetration of market economy discourse into the arts and cultural sector" (1996, 3). In some cases, cultural policy—once viewed almost exclusively as a social priority "belonging" with arts and welfare divisions—has been transformed into *de facto* local economic development policy, in the absence of central state redistributive programs.

Contrasted against the perceived benefits of the arts "as creative economy" discussed above are a range of issues that appear more substantial as comparative research emerges. It is these issues that I am most concerned with in this chapter. I've broadly organised discussion of these under theme headings below: those issues that relate to the context of arts/creative regions strategies; problems with definitions and measurements of "creativity"; and ethical issues about who benefits (and who is marginalised) by policy development.

Contextualising creative arts strategies

As research emerges, it is becoming clearer that place characteristics and the embeddedness of cultural activities in particular locations are incredibly important in determining whether arts-as-creative economy strategies are appropriate, and whether they subsequently succeed or fail (Barnes et al. 2006). Put simply: the context matters. Already existing infrastructures for the arts make an enormous difference, as does the presence (or absence) of local artistic communities, music scenes, writing clubs, local dramatic societies and dance schools (Gibson 2002). The presence (or absence) of arts scenes in neighbouring places also makes a difference. Neighbouring towns with arts scenes enable the possibility of shared resources and "spillovers" of audiences and tourists for other towns (as in Byron Bay in NSW), but in contrast also provide tension and competition, especially if places are meant to increasingly vie with each other for scarce funding.

Also related to this is an understanding of the social and economic effects of transformations of regions to "creative places." By embracing influxes of Florida's "creative professionals," poor regions and parts of cities risk actively displacing lower-income residents via processes of revitalisation and rising rents (see MacLeod 2002; Smith 2002; Ley 2004). In parts of Australian cities, such as inner-west Sydney and West End, Brisbane, and in some rural places like Daylesford in Victoria, "creativity" has had an accelerated impact. The emergence of vibrant arts scenes has made them great places to live, and their reputations grow accordingly; "lifestyle" migration accompanies the rise of service industries, the arts and cafes, and property market rises and subsequent social displacements and exclusions occur (see Curry et al. 2001; Gibson 2002). If places do succeed in attracting the creative class (with commensurate impacts on property markets, tourism and investment), they may well bring monetary benefits, but may also produce unintended social outcomes. Amongst these are the displacement of grass-roots artists, musicians, writers and actors, who themselves struggle to afford increasingly exorbitant rents in those places. In a circular way, the sustainability of arts scenes in places may be damaged by success. Florida himself acknowledges that such displacement is an outcome of "creative" revitalisation, but he does not offer a solution. Care should therefore be taken when presuming that "success" in promoting the arts and creative industries comes without consequences.

What is "creativity" and how can it be measured?

Although the subject of much recent academic and policy discourse, the concept of "creativity" itself remains rather muddy. What is it, and who is "creative"? For arts practitioners, understandings of creativity are normally linked to rather abstract, personal experiences and qualities: inspiration, convergence, hard work, lateral thinking, synthesis, experimentation, daring, reflection. For Florida, analysis of "creativity" is linked to a series of key demographic variables including: numbers of people in given locations that work in the arts; tertiary education qualifications; and "talent" (itself no less opaque). Creativity equates to general demographic characteristics able to be quantified across time and space (see Gibson et al. 2002 for critique and detail on the statistical calculation of creativity indices).

In Australia, the most explicit borrowing of ideas from the "popular" academic work on creativity and the arts was the commissioning of National Economics [1] in 2002 to undertake the *State of the Regions* report for the Australian Local Government Association. The report is actually a series of consultancy projects undertaken by National Economics on an annual basis. Each year demographic and labour-force data are compiled for every region in Australia, enabling local governments to monitor the performance of their locality against others. According to National Economics, it "has become the benchmark report on regional economic development, providing a vehicle for research into the ways regions respond to economic change" (National Economics 2004). Following in the footsteps of the popularity of Florida's book, in 2002 National Economics added to its analysis a computation of the "creativity" of each Australian city and region, in addition to the usual statistics (age distribution, employment by industry etc).

Here, a key text informing local government cultural policy in Australia directly employed Florida's "creativity index," developing statistical correlations between various social characteristics of resident populations, broken down into a "diversity index," a "gay index," a "bohemian index," and so on—to create a map of Australian places (as Florida did for the United States). Results from this ambitious exercise were intended to assist local councils in comparing their area's performance against others, and to predict "the ability of regions to be successful in the modern globalised economy" (National Economics 2002, i). They provided a means of examining the relative "success" or "failure" of places across Australia against the common benchmark of creativity. The report's findings thus enabled observations about the capacity of places to compete with each other.

Following this logic of place-competition for a moment, actual results were hardly surprising. Inner areas of capital cities performed well on the creativity index, and inland divisions fared badly. Of the 64 regions analysed across the country, NSW's North (which includes Tamworth and Moree) was ranked 52nd; Murrumbidgee (including much of the Riverina) ranked 57th; the Murray region ranked 53rd; Far and North West ranked 58th; and the Central West was ranked 51st. "Winning" regions were all in the central areas of state capital cities—and predominantly in Sydney and Melbourne over other cities. The general pattern was that creativity mirrored other, already existing indicators of socio-economic status: poor and remote places fared worse than those where infrastructures and audiences were most developed, and where competition for jobs and property has been most fierce.

Such an exercise might well be useful for locations in gaining a sense of where they sit in the national picture of the arts and creative industries. But it can also be seen as problematic in a number of ways. Most directly in the context of this book, the exercise is flawed by the way in which one tenet of economic rationalist doctrine—place competition—was wholly and unproblematically established as the "common sense" or "naturalised" goal underpinning the analysis. There is a seductive appeal in using broad, measuring techniques like the "creativity index" to support the assumption of compulsory place competition. As with other exercises that map and benchmark regional performance, the certainty in numbers seems indisputable. It helps to explain what Larner (2003, 509) refers to as the "tenacity of neoliberalism" when linked to statistical studies that compare the performance of places in economic terms. The false assumption behind the comparative technique—an even playing field upon which places then compete with each other—is shrouded by the statistics, tables and maps, and there is less need for full justification or discussion of how that very assumption might be problematic. Is the playing-field level? Why must creativity-based arts policy-making necessarily entail that places have to compete against each other? Given the problems related to context discussed above, the point here is that regions differ, have varying capacities to support and promote the arts, and require different sorts of cultural policies that suit local communities and arts practitioners. Many of these may be more effective when geared towards goals of social cohesion, well-being, and belonging than to strictly economic criteria. They also might not require that places compete at all, and rely instead on cooperation and sharing of regional infrastructures, and collaboration on exhibitions, arts programs and touring networks. A very successful example of this latter, more cooperative approach to arts policy is the recent formation of a regional organisation representing community and arts festivals in NSW's Far North Coast region. In this example, festivals and places that might ordinarily compete with each other instead chose to collaborate; their resulting organisation has since been able to share information and research capacity, and lobby as a block for cheaper access to services, advertising and infrastructure.

By contrast, the *State of the Regions* report (and many of the creative city/ regions strategies being developed now) tends to reiterate how places must compete within global markets, the importance of private sector solutions to regional problems, and the inevitability and singularity of the economic development "game" that all places are engaged in against each other. Universal equations of "success" are invoked, and singular "yardsticks" proposed against which places are measured. Thus,

> the requirements of the globalised economy are that all industries, whether high-tech or low-tech, must be best practice knowledge based, have a high capacity for innovation, and be highly entrepreneurial. This is raising the bar for all regions. (National Economics 2002, i-ii)

Much of this resonates with what Don Watson in 2003 called "bureaucratisation of public language," where corporate terms such as "building capacity," "learning and knowledge economies," and "benchmarking"—in this study Australian regions are "benchmarked and linked to comparable US regions" (National Economics 2002, i)—come to constitute a pervasive, but ultimately thin language of strategic policy. Yet the language used here, although clichéd, is not completely empty of meaning. It constructs creativity in particular ways. Creativity is linked to the primacy of global markets (hence Silicon Valley is later described as a "pinnacle achievement") and underlies the capacity of places competing for footloose capital and to attract "creative class" migrants.

Beyond problems with prior assumptions, results of National Economics' creativity index exercise more simply reflect both the limitations of quantifying such an abstract concept, and of applying general statistical theories across diverse places. In this instance the scientific allure of creativity indices actually becomes unstable, particularly when the utility of the benchmarking exercise is, for many regions, severely diminished. To elaborate: in the *State of the Regions* report each Australian region's creative class index score was compared with the US results of Florida's work. The intention here was to enable Australian regions to compare their experiences with "equivalent" places in the United States; hence, according to the summary,

> what is important for an Australian region that ranks lowly on the creative index is to use the information… to see how their comparable US regions are performing. From this group select those that have been doing reasonably well socio-economically and explore what is driving this outcome. (National Economics 2002, i-ii)

The Hunter region in NSW, with an economy traditionally based on steel and coal (and more recently wine production), with an overall 26th position in Australia, would have been 242nd, if measured against the US list, and hence, according to the report, was "most like Cheyenne, Wyoming". The NSW Richmond-Tweed, a boutique sub-tropical region most Australians would regard as highly creative (and thus ranked 16th in Australia), only rated as 207th on the equivalent US list, and was listed as "most like Decatur, Alabama". The industrial coastal city of Wollongong, 28th in the Australian list, was "most like Yuma, Arizona" (Yuma is essentially a desert crossroads town on the Mexican-US border, now reliant on casino tourism attracted by its nearby Native American reservation). In some even more absurd statistical comparisons, Melbourne's inner city was likened to Houston, Texas, (sure to please Victorian readers who consider Melbourne somewhat more 'European' in orientation) and its inner southern suburbs were compared to Reno, Nevada (also of casino fame). Sydney's southern suburbs were "most like" Waco, Texas; its inner-west considered a parallel to Pittsburgh; and middle-ring suburbs equivalent in terms of creativity to Lancaster County, Pennsylvania (a region largely trading on nostalgia tourism connected to its fundamentalist Amish population).

Though the report encourages low-ranking regions to be optimistic about the possibilities of comparing with US locations, learning from them and climbing this scale, the validity and genuine utility of such indexes and comparisons—given the obviously different fortunes facing Australian and American places—is highly questionable.

An ethics of the arts in creative places?

If context matters and creativity is difficult to measure, a more fundamental philosophical problem with uncritical adoption of the creative economy policy discourse relates to the purported social and demographic transformations they are meant to entail. The most common idea, post-Richard Florida, is that places ought to compete with each other to attract members of the creative class as tourists and as residents. This in-migration brings benefits (new incomers, new money, industries, audiences) and opportunities. And in some places (such as Port Kembla in Wollongong, New South Wales) importation of the creative class been proposed as an antidote to social problems of crime, prostitution, unemployment and poverty. (Barnes et al. 2006)

But in this creative economy policy agenda, a dangerous set of social judgments appears to creep into the narrative. In re-orientating policy towards in-migration of a selective social group, some people are targeted as "ideal" citizens (i.e. the creative class), while others are seen as inferior. In Wollongong, it was hoped that the creative class would arrive and trigger the transformation of a "downtrodden" place into a new centre of prosperity; while older migrants, the unemployed, retirees and street sex workers, were positioned as "problem" groups, in that they drained public resources and sullied the reputation of the place. (Barnes et al. 2006) By hoping that the creative class could be imported, in order to reinvent the place, longer-present residents were, by default, seen as "less creative," depicted as redundant or as less capable of being transformed into "ideal" creative citizens.

These kinds of policy discourse contribute to what Haylett in 2003 identified as a damaging conflation of socio-economic inequality and cultural attributes. Observable disparities in income, employment, and education do reflect a social and spatial unevenness produced by capitalist economies, yet moral interpretations of this unevenness are smuggled into the policy agenda, and link disparities to ideas of certain people as "problematic." "Unsuccessful" people are perceived as problematic through a binary opposition to the norms set by successful, creative, quasi-bourgeois society. In Haylett's UK example, "working-classness is often reduced to a condition in need of alleviation" (2003, 56-7). In Australia, "lack of creativity" has likewise become a condition triggering corrective policies. Social constructions of class ("working class," "creative class") have been used to label both individuals and geographical places. Thus, according to the creativity index, whole regions (and the people within them) are depicted as "lagging" and as "lacking" diversity, innovation and creativity. To paraphrase Haylett (2003, 57), target problems (socio-spatial disparity) easily become targeted regions and people—"problem" locations that lack creativity are in need of "hipsterisation strategies" (Gibson and Klocker 2005). In the *State of the Regions* report, the results of the creative index benchmarking exercise are disheartening for those who are struggling the most, that is, those least likely to succeed in place competition against the key metropolitan centres. Individuals and local governments are blamed for their low creativity index ranking (i.e. failure), and are urged to reorient their endeavours towards commodifiable innovation, to become more educated, tolerant and creative, in order to attract more "creative" types to their "lagging" places. Judging some regions as more "creative" than others, in this manner, runs the risk of denigrating not just the political institutions in particular locations, but the people who live and work in those places, casting them as lower (creative) class citizens who must individually bear the blame for their region's lack of performance.

More simply too, there is a limited capacity for all places to compete to attract the same limited demographic group—the creative class. There are only so many young, bohemian, creative types to go around. Creative city strategies just won't "stick" in places that might be better off developing arts policies that speak to already-present populations, whether those populations are diverse or monocultural. In places like Queensland's Gold Coast and the Mid-North Coast of NSW, this may be retirees; in Southeast Melbourne arts policy directions would be better off seeking to include recently-arrived refugees from African backgrounds. In Darwin, in Australia's remote north, it makes far more practical and ethical sense for a creative city strategy to engage with its large and diverse Aboriginal population than for it to attempt to attract *en mass* creative class professionals from southern states. That the focus on creative cities and the creative class has increased the general interest in the demography of places is a good thing. But this interest need not give rise to rather simplistic and haphazard policy directions seeking future transitions in populations that are unrealistic or exclusionary.

Some conclusions

As I stated at the outset, this chapter was not intended as an attempt to de-value the idea of a convergence between arts and economic development policy-making, but rather the direction it seems to be taking. Creativity remains interesting to academics and policy-makers precisely because it implies a departure from norms, being genuinely (even radically) new—an antidote to discipline and restriction. There are clearly many benefits to promoting the arts and creativity in an economic development context: they shift the focus away from large-scale industrial projects and/or export-orientated production (that are invariably ecologically unsustainable); and celebrate elements of social life that bring vitality, vibrancy, surprise and emotional pleasure to a great many people. It is indeed time that creativity, in its various forms, is taken seriously, cherished and rewarded.

What I am most concerned about in the emerging policy direction I have sought to critique here, is the way in which the arts become a means to repeat the usual arguments about market-led development, rather than a way of thinking about different kinds of economic development in places. Instead of presenting alternative ways of imagining futures for places, what seems to be happening is that a singular interpretation of creativity is being incorporated into a rather *uncreative* framework, in which private sector solutions to regional problems, the ideal of creative, independent, entrepreneurial subjects, and the primacy of place competition in global markets all remain paramount ideologies.

Though I have voiced concerns here about current policy directions, much of the "playing out" of the discourses of the arts and creative economy in Australia's cities is still taking place: local councils are still revising and writing their cultural plans, and much is "up in the air" with regard to what changes might actually occur on the ground. Some places will absorb and build on the concepts of creative cities and the creative class, and do very well out of this embrace. Others will develop tokenistic—rather than comprehensive—cultural plans, to comply with state government circulars. These may well be the places that "lag" on the creative index, and which have limited hope to reinvent themselves as "innovation hubs" or centres for the arts (or which might be suspicious of the benefits of doing so). The "creative city" may end up a reality for some, while for others may become a passing fad.

Rather than suggest that the idea of the arts as economic development strategy be abandoned altogether, I would argue that we need to be less slavish to imported pet theories about select demographic groups and place competition. Although outside ideas and inspiration from other places are always beneficial in principle, Australian policy-makers could be more open to ideas that stem from our own places and people. The incorporation of the arts into economic development discourse in diverse and locally-sensitive ways could entail an opening up of new conceptual spaces for policy-makers, that build on, for instance, socially-beneficial elements of policy proscriptions—to enhance community interactions, provide stable incomes and jobs for people, form partnerships and become more tolerant—but without having to dogmatically weld these impulses to commercial and competitive agendas. As Haylett has argued,

> Contemporary policy needs the kind of imagination that can understand something of the texture of poverty and working-class lives as ordinary and extraordinary ways of being. Without such thinking, working-class people and places can only ever be 'less than' those in whose image they are reconstructed. (2003, 70)

Creativity is not an innate biological presence in only some individuals, but a potential element of all our diverse endeavours. We need creativity in arts policy-making much more than overt policy prescriptions about creativity.

Acknowledgement

Parts of this chapter appeared in a different form in an earlier article published in *Geographical Research* (43, 1). I would like here to acknowledge the input of Natascha Klocker, who co-authored that earlier piece.

References

Barnes, Kendall, Gordon Waitt, Nick Gill, and Chris Gibson. 2006. Community and nostalgia in urban revitalisation: a critique of urban village and creative class strategies as remedies for social 'problems'. *Australian Geographer* 37, 3: 335-354.

Commonwealth of Australia. 1994. *Creative nation: Commonwealth cultural policy*. Canberra: Department of Communications and the Arts.

Curry, George, Gina Koczberski, and John Selwood. 2001. Cashing out, cashing in: rural change on the south coast of Western Australia. *Australian Geographer* 32: 109-124.

Florida, Richard. 2002. *The Rise of the Creative Class*. New York: Basic Books.

Gibson, Chris. 2002. Rural transformation and cultural industries: popular music on the New South Wales Far North Coast. *Australian Geographical Studies* 40: 336-356.

Gibson, Chris, Peter Murphy, and Robert Freestone. 2002. Employment and socio-spatial relations in Australia's cultural economy. *Australian Geographer* 33: 173-189.

Gibson, Chris and Shane Homan. 2004. Urban redevelopment, live music and public space: cultural performance and the re-making of Marrickville. *International Journal of Cultural Policy* 10: 69-86.

Gibson, Chris and Natascha Klocker. 2005. The 'cultural turn' in Australian regional economic development discourse: neoliberalising creativity? *Geographical Research* 43: 93-102.

Gibson, Lisanne. 2001. *The Uses of Art*. St Lucia: University of Queensland Press.

Haylett, Chris. 2003. Culture, class and urban policy: reconsidering equality. *Antipode* 35: 55-73.

Landry, Charles. 2001. *The Creative City: A Toolkit for Urban Innovators*. London: Earthscan.

Larner, Wendy. 2003. Neoliberalism? *Environment and Planning D: Society and Space* 21: 509-512.

Ley, David. 2004. Transnational spaces and everyday lives. *Transactions of the Institute of British Geographers*, 29: 151-64.

MacLeod, Gordon, 2002. From Urban Entrepreneurialism to a "Revanchist City"? On the Spatial Injustices of Glasgow's Renaissance. *Antipode* 34, 3: 602-624.

Miller, Sarah. 1996. Development without culture? *RealTime* 15: 3.

National Economics. 2002. *The State of the Regions*. Melbourne: National Economics and the Australian Local Government Association.

—. 2004: http://www.nieir.com.au/about_us/company.asp

NSW Department of Local Government, 2004, Cultural Planning Guidelines.

NSW Ministry for the Arts. 1999. Strategy for the Arts in Western Sydney.

O'Regan, Tom. 2001. Cultural policy, cultural planning and creative industries policy making. In *Convergence, Creative Industries and Civil Society: The New Cultural Policy, ed. Colin Mercer. CULTURELINK* Special Issue, 33-50.

Smith, Neil. 2002. New globalism, new urbanism: gentrification as global urban strategy. *Antipode*, 34, 3: 427-450.

Throsby, David. 2001. Defining the artistic workforce: the Australian experience. *Poetics* 28: 255-271.

Watson, Don. 2003. *Death Sentence: The Decay of Public Language.* Sydney: Knopf.

Notes

[1] National Economics are a consultancy firm in the regional economic development policy area. According to their website,

> The National Institute of Economic and Industry Research (NIEIR), trading as National Economics, is an economic research, forecasting and consultancy group, founded in 1984. National Economics' work is divided into two streams, one financed by subscriptions and products and the other by commissioned consultancy advice, studies and reports. (National Economics 2004)

They are an example of "think tank" consultancy firms increasingly called upon to advise governments on economic policy.

CHAPTER FOUR

MONOLITHIC CULTURAL POLICY

CHRISTOPHER MADDEN

Introduction

When national governments enter into the realm of cultural policy, tensions are inevitable as the monolithic nature of nation grates against the pluralistic nature of culture. Governments throughout history have been eager to recruit culture for their own benefit, and have proved adept at supporting cultures that are agreeable to them, but less disposed toward cultures they view as radical, different, or threatening. There are therefore good reasons to be critical when considering national cultural policies, even those of seemingly benign governments: whose cultures are being supported and why? Whose cultures are being ignored or suppressed and why?

Fortunately there is a healthy scepticism in cultural policy discourse. British cultural policy maverick John Pick, for example, views cultural policy as "high-minded ideals that often [bear] little relationship to the actual actions proposed" (quoted in Schuster 2001, 4). With a similar scepticism, Schuster (2001) identifies three common types of cultural policies: "motherhood and apple pie," where policy is couched in terms too general to be a useful guide to action; "everything but the kitchen sink," where policy tries to satisfy too many priorities to be practicable; and "form over function," where policy is emasculated by a focus on processes rather than on outcomes.

Although scepticism and good governance might dictate that we should know what a government's intentions for culture are, not all governments are willing to spell these intentions out explicitly. When they do, they sometimes use what is called here a "monolithic cultural policy": a single cultural policy document bringing together the branches of the government's cultural involvements under one umbrella—into one, hopefully coherent, whole.

This chapter explores monolithic cultural policies internationally to address the question: if Australia were to have a cultural policy, what should it look like? It surveys monolithic cultural policies in a selection of countries, examines what these policies have in common, what they say and how they say it, and ends with some personal reflections based on this comparative analysis. The approach is neither scientific nor fully comprehensive. The sample of policies is based on availability and it does not necessarily include countries that are a "close fit" to Australia. The chapter outlines personal impressions gained from an international comparison and does not aim to supply a full empirical substantiation for those impressions.

Cultural policy

Any international comparison of cultural policies requires some notion of what cultural policy is. To begin with the trite, the phrase "cultural policy" consists of two distinct concepts–"culture" and "policy"—that are in themselves abstract, complicated and contested, making defining the conjoined term especially difficult.

"Culture" stands for different things to different peoples, communities, societies and nations. A person or an agency can adopt different definitions of culture for different purposes and in different settings. That said, it seems to be becoming standard in cultural policy discourse to consider "the arts" as the heart, or epicentre, of culture or creativity, and consequently of cultural policy (Throsby 2006). In this increasingly common view, the cultural and creative sectors are modelled as a set of concentric circles radiating out from a "creative core," the arts. Problems of defining "the arts" aside, the concentric modelling approach is useful as a mind-map, but it does not always fit well with the cultural sector. The model's hierarchies are blurred by an innate innovativeness of creative activities, and a promiscuous integration—both vertical and horizontal—in the arts and cultural sectors. Consider, for example, how media organisations can distort the model's typology by simultaneously being creators, producers, and distributors of creative content, while at the same time purportedly killing creativity and engaging in activities not related to "the arts" or culture. Furthermore, the cultural centrality of the arts implied by the model has hardly been justified by cultural policy commentators.

What are the mechanisms through which the arts impact on culture? Why might opera be identified as core but not karaoke? Why artists but not religious clerics? The machinery that makes the arts so culturally special, as implied in the "creative core" model, has not been clearly articulated.

Ambiguities in defining art and culture mean that definitions of the arts and cultural policy "domain" are far from clear, and countries therefore adopt different definitions of the domain. Fortunately, the aims and methods for the analysis in this chapter preclude the need to define culture succinctly or accurately. The boundaries of culture can be taken to be "floating"—ie. "self-defined" by the policymakers of each country.

"Policy" is another difficult and highly variegated term. As Torjman suggests, there is no simple answer to the question "what is policy?" (2005, 5) Definitions from public policy theory, like Torjman's, tend to be prohibitively complicated. Dictionary definitions, on the other hand, tend to be oversimplified. The Oxford English Dictionary's definition of policy as "a course or principle of action adopted or proposed by an organization or individual" could cover any manner of institutional actions, including, in the case of a private company, coveting profit.

To set rudimentary boundaries for this chapter, attention will be focussed solely on the cultural policies of *national governments*. Cultural policy should therefore be taken here as "a plan or course of action of national government to influence and determine cultural decisions, actions and other cultural matters."[1] The cultural policies of sub-national and local governments, non-government organisations, transnational organisations, private companies and foundations, will therefore not be subject to analysis.

Schuster's sceptical typology of cultural policies has already been described. In a more neutral vein, a 2006 review of the cultural policy literature undertaken by the International Federation of Arts Councils and Culture Agencies (IFACCA) identifies a number of universal themes in cultural policies around the world (IFACCA 2006). Three major themes are summarised below.

> **Theme 1: Five key elements**
> The "machinery" of cultural policy involves the selection and mixing of five key elements: (i) policy domains that are considered "cultural" (eg. visual arts, performing arts, broadcasting, film); (ii)

policy instruments (eg. subsidy, tax incentives, ownership); (iii) institutional structures (eg. ministry, department, arms length agency); (iv) decision making processes (eg. peer review, bureaucratic decree); and (v) rules and customs that determine the interaction of the above elements.

Theme 2: Multiple vertical and horizontal linkages
Cultural policy involves a multiplicity of policy linkages. Cooperation and collaboration usually needs to be made between institutions responsible for the various policy domains that fall under a government's definition of culture. For example, cultural policy will often involve linkages between government organisations with responsibility for the arts, heritage, broadcasting, and library sectors. Cultural policy also requires external—or "transversal"—linkages to government agencies in other policy domains such as tourism, health, and education.

Theme 3: Complexity
Matarasso and Landry suggest that "the development and management of cultural policy is…one of the most complex areas of modern government." (1999, 7) The themes above have already alluded to an administrative, or mechanistic complexity, and this is reflected in the increasing use in cultural policy of systems analysis, network analysis, and entity-relationship modelling techniques to model cultural policy (for examples see Gray 2000, and Schuster 2003). But cultural policy is not just mechanistically complex, it is also conceptually complex. Culture is an abstract thing, with a multitude of meanings and interpretations depending on who is using the term in what context. Because policy cannot be separated from its conceptual or theoretical underpinnings, cultural policies are also unavoidably conceptually complex or abstract (Volkerling 1996; Weiss 2000; Hugoson 1997).

Invisible, visible and monolithic cultural policies

In light of the themes outlined above, national governments face a daunting task if they choose to be explicit about their cultural policies. Explicitness requires accounting for the multiple entities and relationships in a complex policy system, and it obliges governments to articulate themselves in abstract and often emotive ways. Not surprisingly, many governments choose not to tackle such discomforts. But the paradox is that cultural policies are unavoidable. As Atlas notes, government policies will impact on culture whether they are explicit or implicit:

> Not calling something a policy does not mean there isn't any. Cultural policies… are made all the time. In the United States, policy and policymaking are more often implicit than explicit, and thus they are frequently invisible. (2001, 65)

There has been some debate in the USA about whether cultural policies should be visible or invisible, with academics tending to favour visibility. Schuster, for example, argues that even a poor policy is better than no policy at all:

> Is it not unreasonable to insist that a public agency have a clearly articulated intent, one that is specific enough that we can know whether or not the actions that that agency is taking are in line with that intent? (2001, 4)

And later:

> Whether or not it is considered appropriate to utter the words "cultural policy" in polite company, each society has one, and it ought to be made explicit so that we can all judge how well we are doing. (Schuster 2001, 5)

Others agree. After rehearsing arguments for and against invisible cultural policies, Atlas concludes that the USA government's "policy of not having a cultural policy…prevents the country from being able to have a conversation about the value of art and culture" and that "invisible policies can easily become undemocratic and unaccountable." (2001, 65) Wyszomirski argues that invisibility

> makes it difficult to develop linkages and coordination among agencies. Thus, opportunities both to advance the arts and to improve the social impact of the arts are missed. (1995, 76)

Australia is similar to the USA in that it currently does not have a visible cultural policy. Not surprisingly, David Throsby's recent call for Australia to have a cultural policy echoes the arguments used by his American colleagues:

> By spelling out how the different aspects of cultural policy fit into an overall policy agenda, we can raise the profile of culture in national affairs and provide a clearer direction for policy-making. (Throsby 2006, 33)

A strong case has therefore been advanced by cultural policy experts, sceptics and converts alike, that cultural policies should be made explicit so that a government's cultural activities can be properly evaluated.

A common way of making cultural policy explicit is via a "monolithic" cultural policy: a public document in which a government draws its various cultural interests together into one coherent whole, and that sets out its aspirations for those interests.

A number of countries around the world have a monolithic cultural policy. The great benefit for the comparative policy analyst is that monolithic cultural policies are easily compared, and it was with the aim of making inter-country comparisons that, in preparing for the forum from which this book has sprung, the author went searching internationally for monolithic cultural policies from which to make inferences for Australian cultural policy. The outcome of that search is described in the next section.

Comparing cultural policies

There are now a number of international online databases of cultural policies.[2] The nature, coverage, and quality of the databases vary substantially, and using them can be a somewhat hit-and-miss affair. That said, there is a considerable amount of cultural policy information now available online from which to make international comparisons, and the problem for the comparative analyst is more an excess than a dearth of information. To place some boundaries for the current comparative analysis, search efforts were focussed solely on monolithic cultural policies, as defined, and on policies in English.

Given that the aim of the search was to find information relevant to Australia, the natural first step would be to explore cultural policies from countries that are a "close fit" to Australia (as recommended in IFACCA 2004). There is, however, some problem in this, as countries that might be considered a close fit—such as New Zealand, the UK, the USA and Canada—do not currently have monolithic cultural policies. After a wider search using IFACCA's cultural policies portal, monolithic cultural policies were selected from: Botswana, Ecuador, Guatemala, Jamaica, Mexico, Namibia, and Uganda. These policies form the basis of the comparative analysis here.

The key contents of the policies surveyed are summarized in Table 4.1. Even though the contents and wording of each policy differs, and each has its own distinct "style," the list in table 1 looks surprisingly like a guide to the design of a cultural policy, and to this end may even be worth expanding on further with additional analysis at a more detailed level. Indeed, it was the intention of the empirical analysis to construct a more detailed taxonomy of the contents of cultural policies. But the taxonomic exercise was hijacked by something more compelling: some of the policies were actually interesting!

Table 4.1

Key Elements[3] of Selected Cultural Policies	
General policy area	Contents
Context/preamble	Definitions (culture, arts etc.)
	Map of key agents and stakeholders in the cultural sector and in government
	Issues map of key issues for the policy to address
	Policy rationale that justifies the existence of the policy
Policy framework	Guiding principles
	Vision
	Mission, goals, aims
	Objectives, key priorities
Implementation	Strategies, policies, programmes
	Roles and responsibilities of agencies
Follow-up	Monitoring and evaluation
	Policy review process and timetable

Countries: Botswana (Department of Culture and Youth 2001), Ecuador (Bejarano et al 2002), Guatemala (Ministry of Culture and Sports 2005), Jamaica (Ministry of Education, Youth and Culture 2003), Mexico (Conaculta 2006), Namibia (Ministry of Basic Education, Sport and Culture 2001), and Uganda (Ministry of Gender, Labour and Social Development 2003).

The list in Table 4.1 has a rather dry, textbook feel, and that is how some of the policies read; as if drafted as an exercise in Public Policy and couched in the efficient but deadpan language of missions, aims, objectives, stakeholders, evaluative measures, and so on. However, some of the policies indicate that "dryness" is not a prerequisite of cultural policy. The language used in some policies is hardly vague or mundane; it is uplifting and motivational. The policies seem, to an outsider at least, to be surprisingly bold, interesting, or even radical. A selection of extracts is reproduced below to illustrate.

Jamaica's draft cultural policy (Ministry of Education, Youth and Culture 2003) is spiced up with quotes from Jamaican artists, including Reggae legend Bob Marley, and Hip-Hop-East Coast Rap-Dancehall-Ragga artist Shinehead.[4] Even the policy's title—*Towards Jamaica the Cultural Superstate*—provides a hint that it is something out of the ordinary.

Namibia's cultural policy takes a frank look at the country's past and the role the arts will play in the country's future:

> We should not forget that before independence people were divided and the majority discriminated against on the basis of race and culture... [M]anipulative actions [were taken] to "divide and rule." It served apartheid's purposes to foster a culture of enmity, inequality, isolation and dependence. Certain cultures and values were projected as superior, and a sense of inferiority was conditioned in the minds of many of our people. Cultural actions were therefore also an important part of the resistance to apartheid and colonial rule. Some of our artists showed just what a powerful role art can play in bringing about a change. The African Renaissance, based in arts and culture, must therefore be seen as an important part of our continuing emancipation. (Ministry of Basic Education and Culture 2001, 9-10).

Guatemala's cultural policy has a similar flavour:

> At the turn of the 21st Century, Guatemala has entered a new era. As a result of painstaking efforts, we are at last building peace. We are now prepared to conquer the future, and for that we are laying the foundations for dialogue, mutual understanding and solidarity. It is the time for self-recognition, for strengthening identities. It is the time of being ourselves. (Ministry of Culture and Sports Guatemala 2005, 2)

Both policies have a clear shelf life. The impacts of Namibia's apartheid will diminish as time goes by. Guatemala will not always find itself rebuilding after conflict, nor will the 21st Century forever be a benchmark. But even when outdated, these reflective cultural policies will stand as testament to the character and aspirations of the nation at a point in history. Consider, as a final example, Cuba's current cultural policy:

> [We are] in a world dominated by a fierce neo-liberal globalization, in which a world power is trying to impose its hegemony . . . with messages of mediocre and meaningless consumption, aimed at influencing large sections of the planet's population with a certain specific way of thinking and of behavior, turning such masses into a passive, acritical [spectator]. Therefore, it is ever more critical to cultivate the respect for the national cultural values, from the viewpoint of a deeply humanistic ethics, through which the human being, as critical protagonist and participant, can establish a truly creative dialogue with contemporary culture and face any kind of manipulations. (Ministry of Culture 2005, 1)

One of the policy's objectives is:

> [To make] our people one of the most [cultured] in the world in the next few years as a revolutionary, humanistic and [critical] alternative against spiritual poverty, against globalized skepticism and against an imposed nonsense. (Ministry of Culture 2005, 1)

Balancing constitutionality and flexibility

Another aspect of the monolithic policies consulted that is particularly interesting in considering implications for Australia is how they balance immutable principles with short-to-medium term policy objectives. A number of policies had whole sections that were constitutional in character—sections that set into stone such things as human rights, cultural rights and the rights of creators. International instruments form the constitutional backbone of a number of policies. The United Nations Universal Declaration on Human Rights, which contains references to cultural rights, is a popular reference point. But policies also cite various conventions, declarations and recommendations developed by UNESCO—too many to list here but reproduced at UNESCO (2006)—on issues such as the status of the artist, the protection of tangible and intangible culture, the illicit trade of cultural property, and the prevention of intentional destruction of cultural heritage. Instruments of other international organisations are also cited in a number of the policies, such as those of the International Labour Organisation. The policies surveyed also incorporate a variety of other long-term principles besides international instruments, with clauses referring such things as treaties, bilateral agreements, the nation's own constitution, and relevant domestic laws and statutes.

Important though these "constitutional" elements are, a cultural policy also needs to remain flexible, especially as, according to Atlas, "the power of creativity and the dynamic nature of culture often defy the coherence and consistency expected of policymaking." (2001, 68) A good example of such flexibility is the cultural policy of The Netherlands. The Dutch enshrine flexibility in their cultural policy by having a four-year review cycle built in to their cultural policy process. The latest four-year cycle has seen a change in focus from the "audience development" of previous years, to an emphasis on reducing bureaucratic influence over arts decisions and preventing excessive instrumentalism in cultural policy. In her policy letter, Dutch State Secretary for culture, Medy van der Laan, argues:

> [I]t is time for a new approach to culture. During the past years, [the view has developed] that the cultural sector must be used to solve social issues in the area of integration and emancipation...That has resulted in an instrumentalist policy…The cabinet chooses… reverse reasoning: [it is] not…social awareness in culture [that] must be increased, but rather cultural awareness in society. (van der Laan 2003, 2)

Later, the Minister states:

> Artists and culture creators–both professionals and amateurs–carry out work that has a considerable intrinsic value. That value is not only interesting from the societal, social or economic point of view but also, in itself, more precious than words can express. (van der Laan 2003, 27)

Although periodic reviewing of cultural policy might risk introducing a faddism into cultural policy, it does allow the Dutch to respond to new developments in what is a rapidly changing field.

Implications—the "shoulds" of cultural policy

A number of implications are evident from this unscientific scan of monolithic cultural policies. The aspects of the policies that stood out, that resonated, were those least like the "magisterial" statements "handed down from on high" so maligned by Throsby (2006). And, even though they all tend to have the textbook elements of good policy—combining objectives, instruments, and stakeholders in a rational and logical way—these were not the aspects that were most compelling about them. The survey highlighted the importance of honesty and reflection in cultural policy, and the importance for cultural policy to interest and inspire people. It caused this author to consider a number of over-arching "shoulds" if Australia were ever to develop a cultural policy. These shoulds are summarised below.

A cultural policy *should* be honest and self-reflexive. It should be a "warts and all" reflection of the nation and its cultures. If done honestly, a cultural policy can be a way for nations—and national governments—to face demons, to face up to and right past wrongs, and to articulate things that would not be articulated in any other domain of government.

A cultural policy *should* take an impartial view of culture; it should not be an advocacy document that aggrandizes culture or makes exaggerated claims that culture cannot hope to fulfill. It should recognise that the impacts of culture, and of cultural policies, can be both good and bad, both desirable and undesirable. The culture wars in the USA have illustrated that censorship, in that it is the management of the negative consequences of creative expression (be they perceived or real), is as much a cultural issue for governments as is the promotion of cultural expression. The cultural practices of one group may be abhorred by another group (consider, as an example, female genital mutilation). Policy needs to recognise that not all practices deemed "cultural" by their protagonists are universally celebrated.

A cultural policy *should* be interesting and even inspirational. A boring cultural policy is not worth producing. Furthermore, culture may be an abstract thing, but this does not mean that it should be unclear or vague. Many abstract phenomena can be described with brevity and clarity. $E = mc^2$ is a good example. Gobbledygook and "weasel words" are unnecessary in cultural policy.

A cultural policy *should* be comprehensive. If it overlooks cultural issues, it will not just be incomplete, it will also be weakened, especially if transversal policy alliances go unnoticed. Comprehensiveness should be measured across a number of dimensions. The policy should involve as many cultural "domains" as is practicable or acceptable. This means not just incorporating conventional cultural domains, such as "The Arts," but also other domains that may be considered cultural by groups outside The

Arts power base. It should consider all the "impacts," "values," or "linkages" associated with these cultural domains (call these "associations" for convenience). Cultural policy has a myriad of policy associations—with health, education, youth, crime, tourism, urban planning, broadcasting, and communications to name a few—most of which can be thought of as falling into three broad types: social, cognitive, and affective associations. A cultural policy that ignores one or more of these types of association will be palpably incomplete. An example of a fairly comprehensive "map" of associations can be found in *The Values Study* produced by the Connecticut Commission on Culture and Tourism. (2004, 15) In addition, as current debate demonstrates, a cultural policy should attempt to understand and articulate how these associations relate both "intrinsically" and "instrumentally" to the public interest.

A cultural policy *should* not stop at the "supply-side" of culture. It should not be an industry policy for the sake of supporting the cultural industries. The cultural industries should be supported for grander reasons than their own benefit; if they are to be supported through industry development or employment initiatives, it should be for the benefit of *all* citizens, and policy success or failure measured accordingly.

Finally, the policy *should* be revised periodically. As noted earlier, a cultural policy should enshrine certain long-term principles of a constitutional and human rights nature; principles that outlive political parties and that eschew faddism. But culture and cultural priorities change. A cultural policy needs to move with the times.

Australian Cultural Policy

The last time Australia had anything resembling a monolithic cultural policy was with 1994's *Creative Nation*, the cultural policy of the Keating Labor Government of 1991 to 1996. As Australia's first and last monolithic cultural policy, *Creative Nation* is an anomaly in the history of Australian cultural policy. Historically, Federal Australian governments have relied on an invisible approach to cultural policy.

Since Creative Nation was abandoned in 1996 by the incoming Coalition Government, there have been a number of reviews and policy initiatives for sub-components of government cultural policy. There have, for example, been major policy reviews for the contemporary visual arts and for the major performing arts, both of which have on the whole been welcomed (Department of Communications, Information Technology and the Arts 1999 and 2002). Australia's current cultural policy problem lies not in whether these initiatives were in themselves good or bad, but in the piecemeal approach and lack of strategic coordination that has surrounded them. There has been no stated coordinating policy umbrella for culture *as a whole* that has set priorities and specified the appropriate sequencing of policy reforms for components of culture. Each review has been undertaken in and for itself.

The benefits of such invisible cultural policymaking are obvious. The government enjoys greater flexibility in deciding when and which policy sector to reform, and therefore has greater power to target any kudos or electoral benefits that result from reforms. But there are also compelling benefits to the cultural sector and to society of making the sequencing of such sectoral policy reviews more explicit and coordinated. Reducing randomness and transparency is good governance. In such a highly transversal policy arena, there are also clear benefits to mapping all of government's cultural interests. Making a map where none exists is bound to throw up anomalies, surprises and possibilities. Making explicit links between arts and cultural policy and other policy areas would surely also be welcomed by arts and culture agencies in an increasingly instrumental policy environment, as it would facilitate their negotiations with agencies in non-cultural policy areas of government. The cost of building this cultural policy umbrella is the need for intellectual rigour and the obligation for the Australian government and Australia itself to face up to past and current demons.

David Throsby argued for such an explicit approach to cultural policymaking in Australia in his influential 2006 essay, "Does Australia Need a Cultural Policy?" Rather than a "magisterial" cultural policy handed down from on high, Throsby envisages a cultural "accord" between Australia's government and its people, forged out of a national debate or discussion. Although it is unclear exactly what sort of accord is envisaged, it is at least explicit, and this, Throsby argues, is preferable to invisibility

It seems Throsby's essay sparked something of a debate. The question "does Australia need a cultural policy?" featured on national talkback radio, and the topic of Australia's cultural identity was picked up by at least one commercial television channel. A search on Google news in the days directly following the publication of the essay in February 2006 returned 14 articles in Australia's daily newspapers about Throsby's proposal, including reactions and commentary.

And crucial issues surfaced just a few weeks after the debate began. Commentators expressed concern that a Federal cultural policy would lead to excessive government influence over Australian culture. An Australian cultural policy would, therefore, need to enshrine cultural independence and reinforce or perhaps strengthen policy mechanisms such as the "arm's length principle" upon which the Australia Council for the Arts was founded

Broader cultural issues also arose later the same year when two government leaders—the then Prime Minister Howard and Treasurer Costello—chose the topic of culture to mark their government's ten years in power by arguing that Australia's migrants should ascribe to "Australian cultural values."[5] When prompted to articulate these values, David Humphries in The Sydney Morning Herald of 25 February 2006 reported the Prime Minister suggesting "Australia's core set of values flowed from its Anglo-Saxon identity" (the Prime Minister's own ethnicity). The naïveté in this description of the values of a country such as Australia, with an indigenous culture that extends deep into antiquity and with vast waves of past and present migration, only serves to illustrate just how important it is for the country to enter into the kind of debate advocated by Throsby.

The muddle in the broader cultural discussion over "Australian values" might be partly attributed to Australia's woefully inadequate constitution. Singularly uninspiring, the Constitution says next to nothing about the nation's values. Enacted in 1901 as a document of Federation for separate state colonies, the constitution confines itself to machinery of government issues (Commonwealth of Australia 1901). It says little about the rights, values and aspirations of the new nation, or about the rights of Indigenous Australians. In neglecting to articulate broad social and human values, it provides no framework for planning or evaluating social action. As suggested in this chapter, it is unavoidable that any Australian cultural policy would need to tackle a number of such rights issues. This becomes an even greater imperative in a country with such an inadequate constitution.

While the release of Throsby's 2006 essay prompted debate on particular issues in Australian arts policy—on the arm's length principle, on cultural values, and on cultural issues relating to migration and migrant communities—it is odd that issues surrounding indigenous Australian culture were largely ignored. Such disregard is not only negligent, it fuels paradox and confusion. For example: if, as then Prime Minister Howard suggested, Australia's values are Anglo-Saxon, and those not willing to share these values should not migrate to Australia, where does that leave Australia's indigenous people? Should an ultimatum be extrapolated from his comments that Aboriginal Australians should either assimilate to Anglo-Saxon values or leave the country? This was surely not what John Howard intended to argue, but the absurd corollary illustrates that any Australian national government that choses to develop a cultural policy would need a substantially more sophisticated appreciation of such complex ideas as "Australian culture," "Australian values," and, indeed, the very notion of "Australia" itself. The next section will turn very briefly to the intersection between indigenous cultures and Australian cultural policy.

Indigenous cultures and Australian cultural policy

The foremost recommendation from the list of "shoulds" presented earlier in this chapter is that a cultural policy should be honest and self-reflective. The most critical test for an Australian cultural policy in terms of this "honesty" criterion would be in how it approached issues of indigenous cultures, the cultures of Australia's Aboriginal and Torres Strait Islander people.

It has become something of a cliché to declare Australia's indigenous cultures the oldest living in the world.[6] Of course, longevity does not guarantee survival. In addition to the usual vagaries of colonisation, the arrival of white people in Australia was particularly brutal on Australian indigenous communities. Land was annexed *en masse* in 1895 when it was declared *terra nullius* (ie. occupied by no-one prior to the British Crown).[7] Indigenous populations have had a profound lack of access to political voice, being disingenuously denied suffrage in national elections until 1967.[8] Indifference and apathy among successive national governments has delayed reconciliation between Australia's colonisers and colonised. Although one of the first tasks undertaken by Kevin Rudd's Labor Government was to extend a formal apology to Australia's Indigenous people for past wrongdoings—an important first step toward advancing the reconciliation cause—full reconciliation remains a distant aspiration.

Cultural policy and indigenous reconciliation are inseparable. Take the case of land annexation. *Terra nullius* was culturally destructive to many indigenous communities because, as Strelein notes, "[t]he land…is central to the worldview, the spirit and the history of all Indigenous peoples in Australia. The identities, language and relationships of Indigenous peoples all come from the land." (1999, 2) The intimate link between land and indigenous culture is even recognised in modern Australian law. Under the 1993 Native Title Act, indigenous communities can stake a claim to land so long as they can prove they have "traditional laws and customs" that connect them to the land.[9] Culture therefore looms large in both successful and unsuccessful court judgements over land claims. (See, for example, the references to culture, traditions and customs in DuVé 2005, and Department of the Premier and Cabinet 2005).

The symbols, values and attributes of Australia's indigenous people are the cornerstone of what makes Australian culture unique. Yet, today, the use of the symbols and values of indigenous culture by non-indigenous Australia is too often opportunistic and of one-way benefit. Australia cannot expect to exploit indigenous culture without a clear ethical framework forged through proper reconciliation.

It is hardly surprising then that the government responsible for *Creative Nation* also advanced reconciliation in a way that no previous or subsequent government has. Indigenous culture and arts are, for example, integral to the Prime Minister Paul Keating's 1992 "Redfern Speech" on reconciliation with indigenous Australians, which is still revered by proponents of reconciliation today:

> [Aboriginal Australians] have shaped our knowledge of this continent and of ourselves. They have shaped our identity. They are there in the Australian legend. We should never forget—they helped build this nation. And if we have a sense of justice, as well as common sense, we will forge a new partnership.

And later,

> We cannot imagine that the descendants of people whose genius and resilience maintained a culture here through 50,000 years or more, through cataclysmic changes to the climate and environment, and who then survived two centuries of dispossession and abuse, will be denied their place in the modern Australian nation. We cannot imagine that. We cannot imagine that we will fail.

A cultural policy, if Australia were to have one, would need to revisit these issues of culture and indigenous reconciliation. If done well and in full consultation, the policy could even be a critical step in moving the country toward a full national reconciliation.

Summary

This chapter assumed the task of looking at cultural policies around the world to consider what lessons these policies might have for Australia. The policies selected highlight many things that could be in an Australian cultural policy, if Australia were to have one. Most essentially such a policy should say something honest, useful and relevant about Aboriginal and Torres Strait Islander cultures. On top of that, the policy would need to provide a logical coherence to government's various involvements in culture.

But logicality should not beat the life out of a cultural policy. The overseas policies surveyed for this chapter that are most notable are not the ones that follow the public policy textbook. Nor are they full of pleasantries. They are policies that are honest, gritty, and reflective. Reading them suggests that there is no point in having an Australian cultural policy unless it is similarly honest: a policy that reflects the nation, warts and all. Above all, a cultural policy should be interesting. Just like good art, it should excite, move, and motivate people. A good cultural policy should cause people to stop and think.

References

Atlas, Caron. 2001. Cultural Policy: What Is It, Who Makes It, Why Does It Matter? In *Culture Counts: Strategies for a More Vibrant Cultural Life for New York City*. New York: New York Foundation for the Arts.

Australian Museum. 2004. Dreamtime Online: What is Cultural Heritage? Canberra: Australian Museum.

Bejarano, G. N., J. C. Iñiguez, and J. V. Morejón. 2002. *State Cultural Policies 2002-2012: Building a Nation From Cultural Diversity,* (document for debate). Quito: Ministry of Education, Culture, Sports and Recreation Ecuador.

Commonwealth of Australia. 1901. *The Australian Constitution*. Canberra: Parliament of Australia.

—. 1994. *Creative Nation: Commonwealth Cultural Policy*. Canberra: Department of Communications and the Arts.

Conaculta. 2006. *La Cultura en Tus Manos: Programa Nacional de Cultura 2001-2006.* Mexico City: Consejo Nacional para la Cultural y las Artes, Mexico.

Connecticut Commission on Culture and Tourism. 2004. *The Values Study: Rediscovering the Meaning and Value of Arts Participation.* Hartford: Connecticut Commission on Culture and Tourism.

Department of Communications, Information Technology and the Arts. 1999. *Securing the Future – Inquiry into the Major Performing Arts.* Canberra: Commonwealth of Australia.

—. 2002. *Report of the Contemporary Visual Arts and Craft Inquiry.* Canberra: Commonwealth of Australia.

Department of Culture and Youth. 2001. *National Policy On Culture.* Gaborone: Ministry Of Labour And Home Affairs, Botswana.

Office of Native Title. 2005. Ngaanyatjarra Lands Native Title Determination. Government of Western Australia.

DuVé, Merrelle. 2005. Gale on Behalf of the Darug People v Minister for Lands. Melbourne: Indigenous Studies Program, The University of Melbourne.

Flint, David. 2006. Is it Time to Sue Over Mabo? The Foundation of Australia was Untainted by Terra Nullius. *The Australian.* 10 February 2006.

Gray, Clive. 2000. *The Politics of the Arts in Britain.* London: Macmillan Press.

Herald Sun. 2006. Bring on the culture debate. 25 February 2006.

Humphries, David. 2006. Live Here, Be Australian. *The Sydney Morning Herald.* 25 February 2006.

Hugoson, Rolf. 1997. The Rhetoric of Abstract Goals in National Cultural Policies. *International Journal of Cultural Policy* 3, 2: 323-340.

International Federation of Arts Councils and Culture Agencies. 2006. *Independence of Government Arts Funding.*

Jackson, J., and R. Lemieux. 1999. *The Arts and Canada's Cultural Policy.* Ottawa: Political and Social Affairs Division, Library of Parliament.

Keating, Paul. 1992. Australian Launch of the International Year for the World's Indigenous People Speech. Redfern Park, 10 December 1992.

Matarasso, François and Landry, Charles. 1999. Balancing Act: Twenty-one Strategic Dilemmas in Cultural Policy. *Cultural Policies Research and Development Unit Policy Note no. 4.* Strasbourg: Council of Europe Publishing.

Ministry of Culture. 2005. *Cuba's Cultural Policy.* Havana: Ministerio de Cultura Cuba.

Ministry of Culture and Sports Guatemala. 2005. *Políticas Culturales y Deportivas Nacionales*. Guatemala: Ministerio de Cultural y Deportes Guatemala.

Ministry of Education, Youth and Culture. 2003. *The National Cultural Policy of Jamaica: Towards Jamaica the Cultural Superstate.* Kingston: Ministry of Education, Youth and Culture

Ministry of Gender, Labour and Social Development. 2003. *The Republic of Uganda Culture Policy: Fourth Draft (not to be quoted*). Kampala: Ministry of Gender, Labour and Social Development, Uganda.

Ministry of Basic Education, Sport and Culture. 2001. *Unity, Identity and Creativity for Prosperity: Policy on Arts and Culture of the republic of Namibia.* Windhoek: Government of the Republic of Namibia.

National Archives of Australia. 2006. *Governor Bourke's Proclamation 1835 (UK).* Canberra: National Archives of Australia.

Schuster, J. Mark. 2001. Policy and Planning with a Purpose or The Art of Making Choices in Arts Funding. *The Cultural Policy Center at the University of Chicago Working Paper*, Chicago: *The Cultural Policy Center.*

Schuster, J. Mark. 2003. The State Arts Agency as an Element in the Larger State Cultural Policy Context. *Revisioning State Arts Agencies: Symposium Proceedings.* Denver: The Western States Arts Federation, Denver.

Strelein, Lisa. 1999. *Aboriginal land rights in Australia.* Acton: Native Title Research Unit, Australian Institute of Aboriginal and Torres Strait Islander Studies.

Stretton, Pat. 2005. *Aborigines and the Vote.* Kingston: Australian Electoral Commission.

Stroobandt, F., and H. Waege. 2004. International Benchmarking: Are we Comparing Apples to Oranges? Paper presented at *Recent Developments and Applications In Social Research Methodology: Sixth International Conference on Social Science Methodology*, August 16-20. Amsterdam: Amsterdam University.

Throsby, David. 2006. Does Australia Need a Cultural Policy? *Platform Papers* 7, January: 1-54.

Torjman, Sherri. 2005. *What is Policy?* Ottawa: Caledon Institute of Social Policy.

UNESCO. 2006. *Legal Instruments: Culture.* Paris: United Nations Educational, Scientific and Cultural Organisation.

van der Laan, M. 2003. *More than the Sum: Cultural Policy Letter 2004-2007.* Den Haag: Ministry of Education, Culture and Science, The Netherlands.

Volkerling, Michael. 1996. Deconstructing the Difference-Engine: A Theory of Cultural Policy. *International Journal of Cultural Policy* 2(2): 189-212.

Weiss, Janet. 2000. From Research to Social Improvement: Understanding Theories of Intervention. *Nonprofit and Voluntary Sector Quarterly* 29, 1: 81-110.

Wyszomirski, Margaret. 1995. Federal Cultural support: Toward a New Paradigm? *Journal of Arts Management, Law and Society* 25, 1: 69-83.

Notes

[1] Adapted from the definition of policy from www.Answers.com.

[2] See IFACCA's cultural policy portal *cultural policies around the world* for a selection: www.ifacca.org/links/cultural-policies-around-the-world.

[3] Note that the cultural policies consulted do not necessarily contain all elements.

[4] Classification of Shinehead's musical style from www.mp3.com.

[5] At the time of writing, *cultural* values were not a part of the citizenship accession test, only the affirmation of legal "values" such as "to obey the law, "to serve on jury duty when required" and "to vote."

[6] From the view of culture as a complex, interconnected, dynamically evolving system that stretches back to the beginnings of humanity, this can be seen more as a rallying call than as an acute cultural observation. Nevertheless, contemporary estimates suggest that humans had been living in Australia for at least 60,000 years before settlement by Europeans in 1788 (Australian Museum 2004).

[7] For more on *terra nullius*, see National Archives of Australia (2006). Recently it has been argued that the declaration of *terra nullius* is a "judicial fallacy" concocted by historical revisionists, and that Australian land was, instead, annexed (Flint 2006). Outside of legal curiosity, this argument is practically pointless, as it does not diminish the barbarism of land theft, nor lessen the ethical obligation for redress. The impact on indigenous Australians, and the blot on Australia's history, is the same whether the land was declared *terra nullius* or annexed.

[8] The issue of when indigenous Australians were granted the right to vote in Federal elections is a complicated one. A popular misconception is that Australia's Aboriginal people were granted voting rights in a 1967 referendum. This is not the case—the reality is much worse. Until the 1967 referendum, large sections of the indigenous population had been denied the vote through the "scullduggery" of Federal governments (Stretton 2005).

[9] There is thus a distressing "Catch-22" in the Australian land rights process: in order to reclaim their land, communities destroyed by land theft must prove cultural ties to the land whose very theft led to the destruction of those ties. Such proof is onerous for communities splintered by colonisation. It is impossible for communities annihilated by colonisation. These and other links between culture,

land rights and indigenous reconciliation are more deep and complex than can be addressed here.

PART II:

ACCESS, EDUCATION AND INNOVATION

CHAPTER FIVE

ESSENTIAL IDEAS AND IMAGES ABOUT THE POWER OF THE ARTS ON PARTICIPANTS: CREATIVE MAKING AS A SOCIAL PRACTICE

PAMILLE BERG

The title of my chapter contains an implicit question. We can bluntly frame that question as being: "Does participation in the arts or some form of creative making have any *real* power, and does it particularly affect the lives of its makers and those who experience the results?" Or is participation in the arts really just a form of self-obfuscation or self-deception, the dark awareness of which sneaks into our consciousness when we lie awake in the middle of the night?

The answer which my twenty-six years of participation in the commissioning of art within architectural projects suggests is that, if we choose to make it so, creative making can increasingly become a much-needed 'pacemaker' within our society. By pacemaking I mean the role which I think the arts and design can have in variously setting the pace, keeping the pace, keeping regular, keeping going, keeping alive, and turning over ideas and convictions in our individual lives and the lives of our communities.

With respect to pacemaking as 'keeping alive', regardless of each of our political persuasions and our mutually diverse and divergent views at the moment on terrorist action, the war in Iraq, and the world movement among displaced people and refugees, what I think must bind us all at present is the understanding that these are times in which we must now consciously "act out" our understanding of diversity, of pluralism, of tolerance, and of compassion on a daily basis.

To me the word "blowback" is an essential concept in this urgent need for individuals to live a life of action rather than passivity. My use of the term blowback comes from the title of a book by Asian studies scholar Professor Chalmers Johnson, where he reminds us that "blowback" is the name given by the American CIA to the unintended consequences of covert action (Johnson 2003). I use the concept of blowback here to reassert that in our time, we do indeed reap what we sow, and that the way we structure the participation of artists, craftspeople, and other creative makers in the design and construction of our public places now needs to be a conscious social practice of pacemaking for the long-term survival of ideas that count.

If we ask what it is about our contemporary condition that makes that blowback—both intentional and accidental—from the practice of the arts so critical, I believe it is partly encapsulated in a book to which I often turn by MIT scientist Stewart Brand, *The Clock of the Long Now: Time and Responsibility*, where Brand talks about this particular moment in time, saying:

> Civilization is revving itself into a pathologically short attention span. The trend might be coming from the acceleration of technology, the short-horizon perspective of market-driven economics, the next-election perspective of democracies, or the distractions of personal multitasking. All are on the increase…
>
> Velocity itself [has become] the dominating characteristic of the world's quicksilver economy. "We are moving from a world in which the big eat the small", remarked the head of the World Economic Forum, "to a world in which the fast eat the slow"…
>
> Some sort of balancing corrective to the short-sightedness is needed—some mechanism or myth that encourages the long view and the taking of long-term responsibility, where the "long term" is measured at least in centuries…
>
> [That corrective is the idea of what we can call the "Long Now". The "Long Now"] places us where we belong, neither at the end of history nor at the beginning, but in the thick of it. We are not the culmination of history, and we are not start-over revolutionaries; we are in the middle of civilisation's story. The trick is learning how to treat the last ten thousand years as if it were last week, and the next thousand as if it were next week. Such tricks confer advantage. *Culture is where the Long Now operates.* Culture's vast slow-motion dance keeps century and millennium time. Slower than political and economic history, it moves at the pace of language and religion. Culture is the work of whole peoples. (Brand 1999, 2, 3, 38).

This notion of "being in the middle of civilization's story" and learning new, incisive ways to make our culture be a celebration of that fact, rather than a re-run of old ideas, has been voiced by many cultural historians over the past two decades. I also return often to the works of American cultural historian Suzi Gablik, who in her seminal book, *The Re-enchantment of Art*, quotes author David Feinstein:

> We need new myths; we need them urgently and desperately…Times are changing so fast that we cannot afford to stay set in our ways. We need to become exquisitely skilled engineers of change in our mythologies…"

and says:

> There is a need for new forms emphasizing our essential interconnectedness rather than our separateness, forms evoking the feeling of belonging to a larger whole rather than expressing the isolated, alienated self…" (Gablik 1991, 4-5).

Returning to Brand's two phrases: "culture is where the Long Now operates" and "culture is the work of whole peoples"—these, are, I think, the underpinning of the very concept of cultural pacemakers in our time, whether of the creative makers themselves or of their resulting works. His urging to take the long view and long-term responsibility is at heart about ethics: about the question of how we can act ethically and justly in our daily lives in ways which are consistent with "the long now", and not merely with our temporal desires.

Australian contemporary philosopher Peter Singer in his book, *One World: The Ethics of Globalisation*, notes that we can seek the nature of justice:

> …by asking what principles people would choose if they were choosing in conditions that prevented them from knowing what position they themselves would occupy. That is, they must choose without knowing whether they themselves would be rich or poor, a member of the dominant ethnic majority or of an ethnic minority, a religious believer or an atheist, highly skilled or unskilled, and so on.
>
> If we were to apply this method globally rather than for a given society, it would immediately be obvious that one fact about which those making the choice [about the ethical principles by which they would live] should be ignorant is whether they are citizens of a rich nation such as the United States or of a poor nation such as Haiti [or at the moment, Niger in the

midst of its famine]...If [it is] accepted that to choose justly, people must also be ignorant of their citizenship, [this] theory would become a forceful argument for improving the prospects of the worst-off people in the world" (Singer 2002, 10-11).

In order to understand our own Australian place in such a circumstance and the potential power of creative practice, we must remember that nearly all of us reading this book, except for many Indigenous Australians, are among the top six per cent of the world's wealthiest people when seen in global terms. Simply through having a house with a solid roof in which to live (regardless of whether it is one room or many), adequate food three times per day, a clean water supply, access to medical treatment, and access to education for our children, automatically puts us in the top ten per cent of the wealthiest in the world, and for any of us who have been to university or have had some form of specialised training past high school, we approach the 94th percentile, or the top six per cent in wealth.

So in terms of thinking about the arts or creative making as a pacemaker within our society, what are the responsibilities inherent in our existing in the midst of and having at our daily disposal here in Australia that overweening, extraordinary wealth in comparison to the rest of the world? Are all of us free to set whatever pace we choose, to merely do what we like with this remarkable wealth? Peter Singer, in an argument similar to Stewart Brand's "long now" discussion, goes on to say:

> If the group to which we must justify ourselves is the tribe, or the nation, then our morality is likely to be tribal, or nationalistic. If, however, the revolution in communications has created a global audience, then we might feel a need to justify our behaviour to the whole world. This chance creates the material basis for a new ethic that will serve the interests of all those who live on this planet in a way that, despite much rhetoric, no previous ethic has ever done...
>
> The thesis...is that how well we come through the era of globalisation—perhaps whether we come through it at all—will depend on how we respond ethically to the idea that we live in one world. For the rich nations not to take a global ethical viewpoint has long been seriously morally wrong. Now it is also, in the longer term, a danger to their security... (Singer 2002, 14-15).

But what, you may ask, do ethics in the era of globalisation have to do with cultural pacemakers, public art, and the power of art on its participants in relation to Australian culture? Everything, I would say.

To my mind, if we understand that we do indeed reap what we actively sow culturally, and that the blowback of the pace we actively establish culturally is what will help to preserve or help to kill us, part of what is essential is that we continuously invite artists working in public in Australia through their briefs and processes not merely to entertain us or practice their private thoughts in public, but rather to respond to deep ideas, and to interpret and sound the depths of those things which are closest to our heart as communities. If we want to employ the power of contemporary art as a pacemaker for all of us publicly, we must invite artists in their creative making to remind us of the most important things in the ongoing definition in our humane communal existence, such as compassion, generosity, and tolerance.

This "being-reminded-of" rich, complex, clear, potent thinking and feeling is essential in the definition of our public places and their character, as opposed to the mindless "theme-ing" which constitutes a sickness-unto-death in our times. By "theme-ing" I mean the description and detailing of suburban housing estates or urban shopping malls as "Federation" or "Colonial", or the addition of details of physical form such as bastardised finials to the roof-pitches of suburban houses in green-fields developments without any sense of the architectural language from which the finial derived. But will these attitudes, if put into concrete practice in the way we structure our community development programs, our urban design processes, our public art policies, and our state or council funding strategies, actually make a difference in what attracts us or repels us about ourselves, our public spaces, shopping areas, and housing?

Do we have an underutilised resource in Australia that we can tap for this purpose? The answer is, of course, that Australia has in hand, at this moment, a remarkable and extremely underutilised infrastructure of professional artists, craftspeople, designer/makers, and specialist artisans who, according to Professor David Throsby and Virginia Hollister's 2004 economic analyses of the arts, on average make less than a third of other professionals with equivalent training. As a nation we should not be asking whether we can afford the work of artists, craftspeople, and designers or public art as part of our building projects, but rather how we can afford *not* to include it, thereby continuing this insane squandering of the resource represented in these thousands of highly-trained professionals in our midst who are living in more or less polite forms of poverty in comparison to the rest of us.

In economic terms, it's all pretty simple. If we are going to hold our society together, we need to learn how to live well and comfortably with diversity, disjunction, difference, and tremendous change. What are good art and design about? They are by nature different, edgy, challenging, new, irritating, a bit frightening, and often uncomfortable. As international festival director and performer Robyn Archer has always reminded us in her remarkable orchestration of festivals in this country, that is what the arts have always been, where the edge, the *avant garde*, initially surprises, is new, and then slowly drifts towards the centre, where it becomes standard. That process happens again and again, over and over, and the members of communities who live with potent art which regularly appears in their midst slowly become confident about encountering and confronting the change and difference it represents. Who are the people who can, in collaboration with architects, developers, and local government, make this process happen? The very people in the arts who sit in our midst, highly trained, under-resourced, and waiting. Would it make good economic, let alone ethical, sense to get this process going? You bet.

What is it we should be asking artists in this country to do in participating in public projects, according to the ideals quoted above of Brand, Gablik, and Singer? We want them to be challenged by project design briefs to respond to the deep, rich, essential ideas of ourselves and our places rather than superficial "theme-ing"; to increase the apparent content, resonance, complexity and difference in our surroundings rather than sameness; to find ways in their work to respond to and promulgate in all of us an ethical, "long now" approach instead of short-term, short-sighted values; and to consciously craft new mythologies, adjusted images of ourselves, and rich defining communal stories which leave behind the mistakes of the past.

By this notion of "crafting new mythologies" I mean the promulgation of images and ideas of the individual, the family, the community, and of institutions which are not simply about individual gain, the increase of wealth, and the desire for power at all levels of society, but which instead demonstrate complexity and diversity of ways of thinking about what matters: i.e. about the primacy of seeing the individual within her/his role within the community, and of people living in ways at all levels which are based on "having enough" rather than continually "having more".

Is this hard to do? Not really. Will architects, developers, and councils go broke if we ask them to take this on? No, not if we go about it in the right way. During the fourteen years I was one of seven directors of Mitchell/Giurgola & Thorp's large architectural firm in Australia, we included public art commissioning in projects every year, whether major or tiny, and where the project budgets were ample as well as almost non-existent. Our firm didn't go into bankruptcy as a result, and we all went home every night to a roof over our heads and a healthy dinner on the table. We all have huge capacity from the foundation of our standard of wellbeing in this country to make good things happen and an unmistakable ethical responsibility to do so.

Going back to my title given to me about the power of the arts on its participants (which I take to mean its makers, its collaborators, as well as those who experience the making and the result), I would like to briefly describe four examples of what that power can be in very different forms. The first example is the five-year commissioning project for the design and production of the Parliament House Embroidery as part of the commissioning of over two hundred artists and craftspeople for the Parliament House Art Program in Canberra between 1982 and 1988. The story of the Embroidery is remarkable, starting with a visit from a diminutive 70-year-old embroiderer, Dorothy Hyslop, to our architectural office in 1983, who in a wavering voice said that she represented all of the Embroiderers Guilds in Australia, and that over two thousand embroiderers across the country wanted to make an embroidery for the building as a work equivalent to the making and quality of the Bayeaux Tapestry.

Impressed by the Guilds' determination, the Parliament House Construction Authority, on the architect's recommendation, engaged textile artist Kay Lawrence as the commissioned designer for the national embroidery, knowing full well that because she was not an embroiderer herself she would transform traditional stitches into a new language of design and making, and engaged a national embroidery coordinator to move around the country from guild to guild to link and maintain the very high standards of the project. Our architectural teams worked closely with the textile curator at the National Gallery of Australia to design a humidity-controlled case capable of preserving the work for all time, who also taught the guilds how to work to museum conditions in preparing the imported linen base, how to handle the work itself (designed to be worked on two-metre sections in all states and territories) according to best conservation practices, and in the stitch-by-stitch joining of the completed sections into the sixteen-metre-long work.

Examining the influence of the Australian continent on the patterns of settlement in Australia and employing the words of poet Judith Wright and an indigenous elder, the work in its final form was declared by former Director of the National Gallery James Mollison as being one of Australia's finest commissioned "works of art". What was the effect on these women, many over the age of 60, of participating for five years in this colossal effort? A rubbing of the hands together and asking "What's next?" when it was installed, and an annual pilgrimage by hundreds of the embroiderers from across Australia, still going on fifteen years after its completion, to visit the work and the national place they helped to make.

The second example is the Maui Arts & Cultural Center in Kahului on the island of Maui in Hawaii. From 1991 I worked as the public art consultant for the design and construction of the Center, completed in 1995 and now just going into its phase two construction. This A$30 million performing and visual arts centre was initially a mere dream in the mind of Maui's resident population of 90,000, and the community had no council budget, no government grants, and no direct means of developing the needed funds for its design and construction. Through a remarkable process involving everything from corner cake sales to evening cocktail parties at the homes of the wealthy and benefit concerts performed by artists like Willie Nelson, the community pitched in together and raised the $30 million so that local artists and performers would have state-of-the-art facilities to nurture their creativity, and so that performers like Mikhail Baryshnikov could be enticed to come with his touring dance company from New York, spend some days in the sun, give several inspiring workshops to local young dancers and a couple of major performances at the Center, and have a drink in the robust local Korean bars at night!

When I was asked to work with the Center in formulating its Public Art Program, the board knew innately that the contributions of the visual artists, contemporary and traditional crafts, and the artisanship of Maui's diverse culture of Chinese, Japanese, Korean, Indigenous Hawaiian, Tongan, Portuguese and white European immigrants would be important to making the Center "owned" by the whole of the community, but not how that process could occur. In the resulting Art Program we created, only one of the artists had ever had a public commission previously, and the Program's ten major commissions utilised local assets such as the very high-quality skills of indigenous artists and elders, schoolchildren working with local sculptors, physically-challenged artists supported by teams of colleagues and respected senior artists within the island's contemporary arts. Because the available budget was so low and so hard-won, we enhanced the Art Program budget by planning entire building elements, with the blessing of the architect, as the subject of commissioned works.

After months of painstaking research, one of the commissioned works, again the subject of a long and miraculous process, was the commissioning of Kamaka Emmsley, one of only four remaining indigenous craftsmen on the island who still knew the ancient Hawaiian tradition of dry stone masonry wall building (as demonstrated in the 40-foot high ancient stone *heiau* platforms still extant in the islands), to create 175 linear feet of six-foot-high, four-foot-deep dry stone masonry perimeter walls for the Center. With the permission of up-country ranchers and the Environmental Protection Agency, the over 65 tons of special volcanic stone appropriate in Indigenous Hawaiian culture for the walls was hand-collected from pastures on the slopes of the volcano, in part by the young members of the traditional Hawaiian schools of ancient hula and language. Kamaka trained three other indigenous men while constructing his wall during months of handwork on site (while taking leave from his normal work as a backhoe operator, due to the lack of any market for his precious traditional skills).

The resulting wall is not only a permanent, subtle, very beautiful functional element of the Center, but also a subject of daily visits by the island's tourists seeking out with delight those elements that connect with its deep multicultural roots. Does this kind of work have power for its participants? As the institutional participant, the Center has become a pre-eminent showcase of vibrant local and international art on Maui with a devoted community backing, and as the individual artist-participant, Kamaka is now finishing a university degree in graphic design, despite having come from a local community in which university was more or less both culturally and financially out of the question prior to his wall commission.

The third example is a three-year project finished in late 2003, in which I worked with the indigenous women who have owned and run their own art centre for fifty years at the remote indigenous community of Pukatja or Ernabella, three hours by dirt track southwest of Uluru, to design monumental floor rugs for the public foyer of the new $40 million redevelopment of the historic State Library of South Australia on North Terrace in Adelaide, designed by Hassell Pty Ltd and MGT Architects.

The Library has precious drawings in its archives which were produced in the early 1950s by indigenous children at the Ernabella mission school, and during the building's design, the Library Director asked whether the Ernabella artists might be able to produce some form of work for the new Library. Again after months of quiet research, fourteen of the Ernabella artists, ranging in age from late teens to well over 70, were given full professional design orientations and briefs by myself and staff, with continuous Pitjantjatjara translation, to use their wax-resist batik textile technique to produce scale designs for three monumental rugs for the Library's entry.

The rug designs by elder Pantjiti McKenzie selected to proceed were hand-tufted precisely to her design by women artisans in northeastern China employed by the high-quality rug manufacturer, Tai Ping, coordinated by ourselves, and thirty-four members of the Ernabella community came to Adelaide to perform at the Library's opening by Rupert Murdoch and the State Premier. Rather than letting the project end there, with termination of its possibility of future economic and contemporary cultural stimulus for the Ernabella community, we have continued to work *pro bono* closely with Tai Ping to market Ernabella rug designs to architects and interior designers at the international DesignEx shows in Melbourne and Sydney, as well as with Tai Ping in Europe and the USA for the past two years. Certainly the power of this creative making, both for all of us as the lucky recipients and for Ernabella women taking their rightful place as consummate interior designers as well as creative makers within their own culture, is very clear.

I conclude with some comments about the project I have spent the last seven years coordinating, which is the Art Program for the new St. Patrick's Cathedral at Parramatta, designed by Romaldo Giurgola with MGT Architects, which won the Royal Australian Institute of Architects highest national design award for public buildings in 2003. We started with a lean architectural budget and no funding for the work of any artists and craftspeople, but within four years that had been transformed into an Art Program constituting over 22 per cent of the building budget. In this time of state governments arguing over even one or two per cent for-art policies, the Cathedral project with its 22 per cent Art Program shows what can be done when everything from the front entry doors (the work of sculptor Robin Blau) to the pews for 1000 people, designed and fabricated in Tasmania by designer/maker Kevin Perkins and his colleague Garry Bester, is designed and specially made by artists and craftspeople working closely with the architectural team.

What made this possible financially was the outpouring of interest by the very diverse local Diocese population of members from Italian, Greek, Chinese, Korean, Tongan, and other immigrant and established communities. Early in the Cathedral's design, the Diocese of Parramatta annually rented the Members Dining Room at the NSW Parliament House as an honoured venue for an evening on three successive years, invited all parishioners who wanted to attend to come to a fine evening of drinks and a three-course banquet at which we put up a major exhibition each year of the work-in-progress by artists for the Cathedral, and I gave an after-dinner slide lecture sharing with parishioners in detail how that work in progress was being designed and made. The artists were all flown in for the evening each year with roses on their lapels for parishioners to talk to (so many artists are never given the opportunity to talk about their art with outsiders throughout their entire career!), and everyone was encouraged to touch and handle the artists' prototypes and models in the exhibition.

With a simple listing available at each successive banquet of the cost of each of the works of art underway, the result was an outpouring of support and enthusiasm that was unexpected and unprecedented, with both individuals and groups, whether plumbers, taxi drivers, doctors, or Chinese restaurateurs, deciding to fund the individual commissions under way so that additional commissions could be developed for new artists. When we completed the basic Cathedral and most of the commissions sixteen months ago, many parishioners asked the Diocese's Bishop Kevin Manning whether we could keep having the annual "art dinners" anyway because they had become for so many people such a potent occasion for sharing in the creative making of a place of importance.

In conclusion, I think there is unmistakable evidence here that there is considerable power for both artists and participants in making substantial opportunities for our artists' creative making in support of people's communal heart-held beliefs, lives and endeavours. But the power comes through the process of creating the proper conditions and consciously inviting artists to do the most challenging and important work in their lives through commenting on who we are and what we must be. This is all part of us actively taking on that critical task described by Suzi Gablik as "becoming exquisitely skilled engineers of change in our mythologies", of enhancing the "blow-back" from our creative capacity, and of enabling art to fulfil the irreplaceable role which it deserves in any culture, and most certainly our own.

References

Brand, Stewart. 1999. *The Clock of the Long Now: Time and Responsibility*. New York: Basic Books.

Gablik, Suzi. 1991. *The Re-enchantment of Art*. New York: Thames and Hudson.

Johnson, Chalmers. 2002. *Blowback: The Costs and Consequences of American Empire*, London: Time Warner.

Singer, Peter. 2002. *One World: The Ethics of Globalisation*. New Haven: Yale University Press.

Throsby, David and Virginia Hollister. 2004. *Don't Give Up Your Day Job: an economic study of professional artists in Australia*. The Australia Council for the Arts.

CHAPTER SIX

BUILDING INNOVATION: THE IMPACT OF EDUCATION IN AND THROUGH THE ARTS

ANNE BAMFORD

Introduction

The arts resonate with the patterns of nature, the globe, the universe and humanity. Art marks our identity as humans and instils in us a binding sense of community and reciprocity. Through engaging in the arts, children and young people gain the opportunity to use their hands, minds and bodies in satisfying ways. The arts allow us all to explore the non-verbal parts of our minds and develop ways of learning within the context of a busy, fractured world filled with multi-layered symbols and rich meanings.

The arts articulate life concerns. They acknowledge those things that we most care about. Through ritual, the arts embellish our physical, cultural and social environment and allow us to deal with anxieties and uncertainties—becoming necessary keys to accessing the deeper and higher parts of human nature. By engaging with the arts, we learn to appreciate our place in the world and to value our tangible and intangible heritage. This is particularly important in an increasingly diverse and transient society.

Aarons described art as "soul or thought prints in the sands of time" (1991, 36). So what makes the arts different from just another mark animals leave behind, a footprint or discarded feather, as a sign of their existence? The arts are different because they evidence people mentally engaged in the act of conception—conveyed in finite terms as visual marks, sounds or movements. These conceptions exist as the metaphors of both rational and irrational thoughtfulness. In this way, the arts are a cognitive form of communication.

The arts in education are far more than merely what is contained in art galleries or concert halls. The arts are symbolic and universal—yet also highly social and specific. Given this, how can the value of the arts be manifest within a climate of educational change that demands transparency, measurability, rankings and ultimately—perhaps hopefully—quality?

Arts in education – how we currently stand

Education is expected to achieve an ever-expanding range of social, aesthetic, economic and cultural outcomes. Teachers complain that they are inundated by new and often competing initiatives and policies. National or state education policy frameworks are increasingly being informed by ideas, personnel, principles and strategies often derived well beyond the context in which they are to be administered.

There is a growing sense of accountability in teaching of the arts and closer monitoring of perceived quality through seemingly narrower standardised tests and assessable outcomes. The removal of decision-making powers of the teacher has led to a loss of confidence in—and within—teachers, and increasing classroom pressures seem to echo the social pressures in the world. Combine this with a generally poorly paid and aging teaching population, and education is faced with a situation that makes policy implementation an increasingly difficult task.

This is further exacerbated in the arts where there is a general perception that the *lip service* given to the value of arts and culture within education policy is not matched by perceived education priorities or by quality classroom practices at the coalface. The result is that most children in the world receive either no arts education or superficial, poor quality provisions (Bamford 2006).

Despite this, in recent times there has been a growing interest in the arts as a way to reconstruct curriculum and to improve the quality of teaching and learning. Driven by countries such as Finland, the United Kingdom, Australia, New Zealand, Colombia and Cuba, systems of education are realising that the arts offer enormous potential to rejuvenate educational standards. Ironically many of the economically developing countries have followed blindly the unsuccessful models of western education and moved away from cultural practices and arts mentoring within traditional learning. Countries such as India, Korea, Mongolia and others have realised that their creative futures depend on the sorts of time-honoured arts learning that has been part of their culture for many years (Bamford, 2006).

International debates concerning the formation of national education policy priorities and practices tend to focus on very narrow standards measures and fail to present visionary responses to a very rapidly changing world and learning environment.

In 2004-5 UNESCO, in collaboration with the Australia Council for the Arts and the International Federation of Arts Councils and Culture Agencies, commissioned research to establish the impact (if any) of arts-rich programmes on the education of children and young people around the world.

The findings indicated that the arts make a valuable contribution to the total education of children especially in relation to academic performance, well-being, attitudes to school and perceptions of learning but this impact only occurs where there is high quality programmes (Bamford 2006).

The global research asked countries to submit up to three research-based studies conducted in a given country in the last 10 years (i.e. since 1995). The process did not stipulate which studies should be submitted and relied instead on the countries themselves to make curatorial quality judgements about which research was of most value to submit. The author and research team did not inflict research bias upon this survey by "adding" other studies we may have felt were valuable—but were not put forward by any country. The purpose was to identify the research which was underpinning countries' policies, not to—as the report clearly states—produce an anthology of all research in the field. This process netted over 250 empirical research studies from all over the world.

Through the research, it was apparent that within arts education there are two distinct areas: namely, Education *in* the arts and Education *through* the arts. The former can be described as being sustained and systematic learning in the skills, ways of thinking and presentation of each of the art forms—dancing, visual arts, music, drama and others—producing impacts in terms of improved attitudes to school and learning, enhanced cultural identity and sense of personal satisfaction and wellbeing. Concurrently, Education *through* the arts—that is, education which uses creative and artistic pedagogies to teach all curricula—enhances overall academic attainment, reduces school disaffection and promotes positive cognitive transfer. These benefits are only accrued where there were provisions of quality programs. Poor quality programs were seen to actively inhibit benefit (Bamford, 2006).

While there are a number of benefits associated with high quality arts-rich education, there are specific areas that can be highlighted in terms of future innovations and the value of the arts within education. In particular, the benefits of enhancing communication skills and critical thinking and greater connectivity of learning experiences seem to be at the heart of building sustainable education futures. Concurrently, quality arts education has a number of key social impacts—such as increased confidence, greater collaboration, feelings of well-being and happiness which both directly and indirectly impact on educational quality. These conditions exist within an educational climate that frequently struggles to keep pace with changes generated by technology and the increasing realisation that future education needs to change to meet the learning and economic needs of young people.

To achieve these sorts of benefits, the rhetoric of policy that supports the inclusion of arts education within the total educational experiences of the child needs to be backed by substantial implementation and monitoring structures that ensure children receive high quality programs. Quality arts programs afford the opportunity to initiate sustained educational reform and greatly enhance the overall excellence of education.

The Arts and Learning

The arts give us a renewed sense of connectedness and relatedness with each other and our surroundings and contribute directly to intellectual development. Through the arts we develop spatial awareness, cognitive sorting, the ability to process material from a range of sources into meaningful categories, metaphor and analogy, changeability and adaptation. (Bamford 2006)

While certainly the area of learning modalities and the impact of the arts requires far more research, there are clear indications that the arts promote a different form of learning and are effective across a range of individual learning styles. Much of the literature and policy documents referring to the value of the arts to learning focus on the use of terms that are aligned more to activity processes, than learning outcomes. Words such as "exposure," "experiential," "expressive," and "explorative," (Bamford, 2006) appear frequently within both policy documents and case studies reporting the impact of the arts. These terms can seem somewhat "elusive" objectives but such "e" words hint at areas that may be seen as being core values in arts learning.

While these supposedly "soft" objectives have often been given less value within the learning process due to the perception that they are difficult to observe or measure, research suggests that a broad range of learning competencies occur with frequency—and in greatest concentrations—within quality arts education programs (Bamford 2006). Interestingly, research being undertaken in fields as diverse as justice, health, politics, informatics and the sciences stress these types of "e" learning may be the most relevant for building an innovative, resilient and sustainable world knowledge and service economy. Graduate tracking studies suggest that individuals that possess a rich source of these learning processes are very employable and effective (Anderson 1998).

While they may be more elusive objectives to measure, their relevance is paramount and efforts need to be made within education to extend the boundaries of the measurement systems to capture the core learning that occurs with quality educational provisions. Effective models of arts-rich education are perhaps the best public demonstration of holistic and innovative approaches to learning and can be used to both reveal and substantiate the creative energies of children and their teachers.

This is supported by the growing body of verification showing strong causal links between an arts-rich education and higher levels of academic achievement. The extensive collection of world research case studies initiated by UNESCO revealed evidence of enhancement in literacy, mathematical and language learning in such contexts (Bamford 2006). The studies from over 60 countries showed that both education in and through the arts improved students' confidence to learn and developed problem-solving strategies, critical thinking and collaborative learning. The research studies contained within the compendium demonstrated that students involved in the arts as part of their total educational/school experience performed at a higher academic level and were more community minded than the "low arts" group.

Schools with strong arts programs were perceived to be more dynamic, interactive, energetic and innovative than non-participating schools. Consistent involvement in the arts showed up pronounced academic advantages.

The arts in schools also served to rectify the inequities inherent in education. The more affluent you are the more you tend to engage in the arts, but in lower socioeconomic schools where the arts were introduced, the benefits were immediate and the arts served to in some way redress the balance between socioeconomic groups.

Critical thinking

In the wake of the rapid expansion in different communication modes and the changes in technology, the need for people to be fluent in the arts, as an essential form of meaning-making, also grows. Our schools need to focus on arts-rich education that encourages critical thinking, problem solving and reflection. The new reality is that education for the future should foster critical and aesthetic realms of learning (Scott 1988).

The arts represent a very practical display of thinking and feeling. Through critical study of the arts people construct meaning in a unified form (Walker 1996). The arts are a valuable partner in the development of reflective ways of thinking and learning. To formulate concepts through the arts is to engage thinking that involves elaboration, fluency, originality, focused perception, imagination, multiple perspectives, layering, intuition and logic.

Interestingly, we use the term "art" when describing something of far superior quality, such as the "art of teaching," the "art of public speaking," even "the art of financial dealing." Art—above all—is an act of intelligence, whether overt, covert or unconscious, reflective or reflexive.

Through the arts, students construct meaning into a unified form. The arts are rich in connections. The arts emphasise how phenomena relate to each other. Holistic thinking and synthesis of ideas are fundamental to the future world. Given this view, art becomes a basic content area in school curricula. The arts are fundamental to education as a means to provide students with the social and intellectual survival skills for the unforeseeable future (Hicks 1993).

Schools need to focus on education that encourages thinking, association, problem solving—and the creation of problems needing to be solved. Future values are embedded within the critical and aesthetic realm of learning. It can be argued that the arts assist in producing critical thinkers who are open-minded, can interpret visual evidence, respect clarity and precision and look at different points of view. To think critically, a person must apply levels of reasoning that go beyond what is obvious, stated or recounted. In this way, the arts have the potential of developing critical thinking by providing a way to involve young people in identifying, evaluating, and constructing arguments, using a variety of forms that extend beyond written or verbal communication. It is this ability to call into play multiple communication modes that is at the heart of contemporary technology and communication systems. The interplay between arts and technology means that aspects such as transformation, fragmentation, synthesis and abstraction have become fundamental literacies for children. The arts bind these new literacies in a way that is connected and holistic.

Connectedness

Learning in the arts is multi-dimensional. It involves making connections, playing with ideas, extrapolating, evaluating and criticising. It is highly co-operative and collaborative. It would also appear that learning in the arts plays a more general critical role in the development of other disciplines.

Through the arts, youth construct meaning into a unified form. The arts emphasise how phenomena relate to each other. Holistic thinking and synthesis of ideas are fundamental to the future world. Given this view, the arts are core to education as a means to provide students with the social and intellectual survival skills for an unpredictable future.

Transdisciplinary learning and exposure to people working on the edges of disciplines are at the heart of future learning capabilities. Transdisciplinary learning includes both a combination of interdisciplinary and multidisciplinary learning approaches. This means that it is a model where learning occurs simultaneously within, across and beyond discipline boundaries. In practice transdisciplinary learning in—and through—the arts involves learning fully within the arts, involving the arts to learn in other forms of inquiry, and then importantly, opening-up spaces to allow new thoughts to emerge beyond the conceptions of single disciplines.

While it seems obvious to say that education should be about inquiry into the unknown, schooling frequently falls short of challenging students to ask questions—being content with students answering questions rather than seeking conundrums.

While not happening in the majority of mainstream classrooms, transdisciplinary trends should be increasing as "singular content" and "classroom isolation" are seen to be inadequate to meet the needs of a global society (Hicks 1993). To achieve greater innovative use of transdisciplinary strategies in classrooms, closer relationships should be forged between the arts and science, mathematics, engineering and manufacturing. This is difficult to achieve within educational structures—both internal and external—that have rigid arrangements of time and impermeable boundaries around areas of learning.

The arts ask questions—often unanswerable questions—in a way that challenges students to move beyond comfortable positions and pre-existing knowledge. Through the arts, students are exposed to situations that build aspects such as courage, daring, risk-taking, speculation, flights of metaphor and other activities and characteristics that are needed to operate beyond the known. In a world of rapid change and constant innovation, the ability to fly beyond the limits of the known is a vital component of education.

In the last century, education was not complete until a young person had travelled, and teachers tried to bring real world examples and experiences into the classroom. At least in part this was in recognition of the need to move a learner beyond the tendency to think that "the way it is done here is the only way" and that others are just a little "weird" if they do it differently. Without wanting to extend the modernist image of the artist as struggling eccentric, it is nevertheless the case that providing alternative ways of thinking and different models of what it means to learn—and what is valued as learning—can be very empowering within education. The ability to reflect on one's learning and to reassess continually the effectiveness of different approaches to learning is of vital importance in a future driven by consumerism, technology and increasingly smaller and more collaborative units of production.

New technology

The growing importance of technology has led to a renaissance of interest in the arts. "User friendliness" is now synonymous with aesthetics and as our reliance/use of technology increases, it is essential that the arts lead the way in this aestheticisation of communication. The media is now what society uses to make its mark and the forms it takes are now largely not written or numerical but music and images.

Today the arts are about music, image, fashion, architecture and communication. People take a picture with a filmless camera, to be stored and perhaps compared to another visual experience in the future—one that has not yet been decided. There is a switching of roles. *I* am the artist. *I* am at once the producer and the consumer of my own media forms. *I* can control how, when and whether I will consume any, all or part of the experience. All children, no matter what career or way of life they choose, need to develop understanding about the impact of the mass media in their lives (Loveless 1990, 202).

The role of the artist and of the scientist both evolved out of the tradition of the "craftsman." Each looks inside to identify new questions and new ways to see things and then each uses design as the vehicle for clarifying those perceptions, seeking various forms and finding processes to communicate that view. As Loveless notes when talking of new technology:

> The primal technology of sticks became that of stories, and then of charcoal and raw pigment. In time the berries yielded to chemicals, the slate to film, paper replaced the walls of caves and fire gave way to electricity. The still image was set in motion and became moving pictures. Then came video and video married computer and all three visual marking systems became a moving picture. And so for the twentieth century child, alphabet soup has been replaced with apples and PCs. Flash cards yielded to calculators. Sounds turned into recordings. Still images into TV, computer graphics and satellite communication networks. (Loveless 1990, 202)

Teachers need to exploit the perceptual and cognitive aspects of the arts in their preparation of children for the future. Aesthetics in the Information Age is about a greater focus on process. It must extend students' cognitive vocabulary. Programs should stress openness to collaboration and cooperation with a greater focus on spontaneity and the unexpected.

Communication

The purpose of the arts—at least in part—is the conveyance of shared, culturally significant meanings. This communication is skilfully manipulated through sensuous mediums. There are many modes of communication, including linguistic, visual, audio, gestural and spatial. The arts use all these forms interchangeably and simultaneously.

The arts as a language emerged at the time of cave dwellers. The current age of information and communication technologies brings with it a renewed consciousness about the communicative role of the arts. There has been a sharp increase in the use of the arts in all levels of multi-communication systems (Hicks 1993).

The arts exist now as the natural and cultural think-prints of our time. It is interesting, that when writing was invented it did not replace the existing art forms. On the contrary, the more advanced our communication has become, the more significant and vital the various forms of the arts are.

Art is about interpretation, confusion, multiplicity, relativity, ephemera, fragmentation, chaos and disjuncture. Teachers are realising this and many will now accept videos, computer images, performances or visual presentations as better than a written paper. As students have become involved in desktop publishing, they have readily incorporated design and aesthetics. They have learnt that the space on a page is as important as the copy. Effective "whole language" instructional programs encompass visual, sound and dramatic elements.

Modern communication has become more insidious. Students who have not been taught to appreciate the arts from a critical standpoint remain ignorant to sub-surface meanings and allusions. Students need to be aware of visual symbols, iconography and the complexity of communication in a global era.

Our communication incorporates multi-modes of delivery and form. To read these complex forms, training is most effectively done through the arts, where multiple forms have always existed. "Christ on the Virgin's lap is not a child, but a numinous representation of the glory, majesty and dignity of Christ the man." (Read 1946, 30).

The arts communicate through the senses to instil values. The arts are a human activity consisting in "that one man consciously, by means of certain external signs, hands on to others feelings he has lived through, and that others are infected by these feelings and also experience them" (Read 1946, 219). The tragedy of our schooling at the moment is that education systems are largely ignoring this kind of literacy.

We expect written literacy and numeracy in our schools, and see these as vital to the future employability and adulthood of our children. Yet very little importance is given to aesthetic literacy. The arts have always been a fundamental form of communication. The future is an era dominated by the arts as multi-dimensional communication, and yet schools are not seeing this as vital in the curriculum. The prevailing attitude is that the arts are nice but not really necessary!

Economic impact—an age of aesthetics

In our economy today, innovation is necessary for survival. Innovation demands free flowing ideas, which in turn requires people who are creatively educated. The business world has made the art world its partner. Companies focus their attention on selling aesthetic environments rather than products. For example, many restaurants are successful because of their ambience, as much as the quality of their food.

But, if we have entered an age of aesthetics in society at large, how can we ignore aesthetic education as part of education?

Aesthetics earns money and form is now more important than function. Taking good quality as a given, consumers now want an image, a lifestyle, a spiritual encounter—put simply, they want an aesthetic experience. Products they buy, things they do and their education must all make personal aesthetic statements. I would like to relate an example described by Hicks:

> Nine years ago a small furniture dealership had a staff of eight sales and secretarial people. Five years later the company had a staff of sixteen. Seven of the eight new employees were artists. The senior executives of the company did not anticipate the need for artists, but all of a sudden they were there. What happened to create this change? One key factor made the difference. In five years the company went from selling furniture to selling aesthetic environments. The customers had reached a point where the furniture and the quality of the furniture were "givens." A given means a

> phenomenon or object is accepted automatically—no longer of concern. Holistic attention was focused on environmental beauty and personalization rather than function alone. The company had gone from selling parts to selling wholes. The bottom line was competition and profits. To be competitive the company had moved from an "art is nice but not essential" rationale to an "art and aesthetics are essential" rationale. Artists became important as salespeople as well as artists. Artists now made the initial sales: more traditional sales types made the wrap up sales. The old stereotype about business being outside the domains of the arts had been shattered. (Hicks 1993, 45)

While the opportunities for employment within the creative sector continue to grow and the so-named *creative industries* are predicted to be a major driver of economic success, the value of arts education should not be conceived solely as vocational education.

Firstly, while the size of this sector has increased, the number of people employed has not increased at the same rate. Employment in this sector is largely characterised by individual or small business activities (Throsby and Hollister 2003). Given this, the value of the arts in supporting the creative industries may be less about building a capable population and more about resilience, collaboration and ingenuity.

Secondly, growth in the creative industries is driven by creative consumers—not by a growth in creative workers. In this regard, arts education may have the most to offer in terms of developing widespread appreciation for the arts and a "palette" for diverse arts options and audience development. In this way the creative industries are a consumer—not producer—driven industry. Given this, the value of the arts may be more aligned to values of consumer education and media studies.

If we look at the growth of ICT-based cultural industries, it is evident that the consumers of these are also responsible for the generation of creative product and content. For example, it is the youth market from China who play computer games who are now the most active producers of computer games. Similarly, the development of podcasts, the success of YouTube and the like, indicate that future creative industries are likely to be reciprocal and the gulf between producer and consumer will become almost invisible. This challenges creative professions as more people move into the realms once owned by the industry. For example, products such as Photoshop were previously the domain of professional designers, now anyone can readily use this technology and create their own designs.

Whereas media such as TV, film, fine arts and theatre have traditionally been produced by a small number of professionals to be consumed by a large number of non-arts people, the future trend is for an equal balance, with people who make arts also being the consumers of such art and vice versa. This further underscores the value of arts education as core for all students.

The shift between knowledge consumers and knowledge makers seems to be most profound in the arts. We have not, for example, seen an emergence of science or maths "in the home market" to the same extent. This will lead to greater decentralisation of the arts. In this participatory world, it will not just be the cultural capitals that will have a monopoly of creative production and exhibition.

Social impact—the necessary behaviour

The arts are an intrinsic part of the way humans operate in the world. The arts are not a "thing"—but rather a human behaviour that is the finest example of what it means to be a person. Children are born aesthetically aware and engage in the arts long before they can speak or write. People practise the arts all over the world. The arts have always been part of humanity's most vital concerns and all societies engage actively in the arts. In contrast to becoming an extinct curiosity, the arts are flourishing in contemporary global society and we all rely upon their powers to get our messages across. In short, we have evolved to be "arts-inclined" individuals.

The arts are not a dispensable luxury or ornament—but rather an essential part of our nature. The universality of the arts would suggest that an important human appetite is being satisfied. If you watch young children at play, you will see them naturally communicating in artistic forms. The arts ritualise and stylise aspects of life, making them universally understandable and easier to deal with. Art is a normal and necessary behaviour of human beings that, like talking, working, exercising, socialising, learning, loving, caring and playing, should be recognised, encouraged and developed in everyone (Dissanayake 1999).

Cultural planning enables people to see themselves as creators of culture. Participants in the arts have stronger networking abilities, a robust sense of identification, and feelings of belonging (Bamford 2006). The arts allow for the reorganisation of knowledge and the grounding of this knowledge within an individual's tangible and intangible heritage. This is a particularly important impact of the arts in a world climate of physical and cultural relocation. Increased globalisation means that cultural identity has become an assemblage, and the arts are used as the detritus of this assembling process. Intersubjective cultural exchanges become the building blocks of identity and locate us within the world. Without substantial senses of position—and perhaps playfulness within this positioning—the power of knowledge cannot be fully realised. At the core of future knowledge economies is the recognition of culturally diverse knowledges we hold and the way the arts can be used to both activate and enact these.

Intrinsic value—the aesthetic sense

While there is a growing appreciation of the societal and critical implications of the arts—including in Jacques Ranciére's work (2004) which acknowledges the value of art not only as a way to represent society but, importantly, as a way to reshape it—it is important to stress that the arts cannot be reduced to mere functionality and material benefits.

A work of art has an intrinsic value, not in the sense of what used to be known as *l'art pour l'art,* but because creative expressions liberate us and make us more appreciative of life itself. This, of course, is not a novel, let alone an original discovery, but it is one that is often overlooked. At a time when many things are measured and reduced to tables and numbers, it is worth bearing in mind that "through a work of art a truth is experienced that we cannot attain in any other way" (Gadamer 2004). A quality arts education is one that facilitates this experience—and hence leads to other benefits.

It is easy to be enveloped by the functional value of the arts and overlook the added joy and pleasure the arts bring to our lives. Aesthetics is about the culturally based "rules" that govern what we define as being beautiful. An aesthetic sense is inherent in all people. We take pleasure from beautiful things.

We know very little of the mechanisms that govern our taste in beauty, our imagination or our creativity. We do know however, that the arts and beauty form a wondrous daydream in which we all engage. As you experience a new piece of music, art or dance you are moved by a deep sense of awe—awe at being in the presence of a manifestation of the act of creativity.

The arts are an expression of the ideal in a form that can be understood by others. Beauty can manifest itself in many ways. The structure of art is not always obvious and beauty may be illusionary or subtle. Yet in a perfect creative artwork or performance, all the elements are interrelated: they cohere to a form of unity that has value greater than the mere sum of these elements (Read 1946, 65). Every work of art has some principle of order or lucid structure. These processes combine to stimulate us in such a way as to bind interest and attention, that is, the arts simply fascinate us (Dissanayake 1999). The grouping of sounds, images or words becomes a sense of "rightness." The elements are positioned in such a way as to ignite our senses and we respond to this inherent sense of order.

Arts in the schools of the future

Aligned with this, school reform requires the reconfiguration of schools to a more aesthetic order. Schools of the future should be less bound by the physical walls of the school and become centres for connecting the child with a range of learning resources.

The Internet has fundamentally changed the way we communicate and access knowledge. Students can readily access information from galleries, museums and the broader educational community. In accordance with this trend, schools should extend the classroom boundaries to include art and cultural institutions and be flexible with timetables and structures to allow students to become immersed in their investigations.

What implications does all this have for the future of education? Our schools must be places where we are challenged to dream about the future. The young people of today will be the inventors of the new cultural patterns and social philosophies of tomorrow. They will invent/design the materials, conditions and community for the future world. Students need to be provided with sustained and sequential learning both within and across the arts.

The arts are an excellent example of lifelong learning and students should be encouraged to see aesthetics as an integral part of their total learning. The need for re-education is a growing trend because of rapid social and technological change. The workers of the future will be more creative so our education system must emphasise problem-solving skills, perceptual development and evaluative skills.

In addition to these demonstrable skills, the arts have an enormous part to play in the total experience of education. Students develop greatly in terms of risk-taking, task persistence and ownership of learning. The arts promote a shared spirit and encourage the student's motivation to learn. Through establishing connections with students, the arts offer something unique and intrinsic to the educational community.

It is important that teachers are well equipped for this predicted future of education. Teachers who see themselves as uncreative are major stumbling blocks to making the arts central to quality education. As products of a system that did not include the arts and was not centred on risk-taking and collaboration, many teachers find it hard to make the sorts of leap needed to envision future education. Even well-meaning teachers often cite their own lack of artistic talent as the reason they cannot engage their students in the arts (Anderson 1998; Ashton 1998; Chia 1995). Furthermore, the arts are rarely given serious consideration in either a teacher's pre-service or in-service education (Bamford 2006).

A caution—the dangers of impact measurement

Throughout this chapter, the emphasis has been on the impacts of the arts within education, but it is important to caution against an overemphasis on impact measurement. To focus only on what art "does"—its function—could be detrimental to the further development of arts education.

If we compare the arts with other curriculum areas such as mathematics and science, you notice that these disciplines' collective energy is not spent justifying their existence. For example there are not debates about whether mathematics is relevant in the future nor attempts by mathematics educators to claim that studying maths improves school attendance, builds literacy or makes children more confident! For such disciplines, it is a given that they are a core part of education. This acknowledgement allows these disciplines to focus more on pedagogy, theorising learning and researching methods of assessment. The arts must *see themselves* as being core and move away from constantly rationalising their existence.

While this chapter has highlighted some impacts of the arts and the relevance of the arts to future education, one should be watchful against the inappropriate use of impact measurement and the dangers inherent in reductionist models of measurement. As Marion Pastor Roces cautioned at 2003's 2nd World Summit on the Arts and Culture,

> Instead of asking how art can or might intersect the discourse between development and culture, it may be the wiser question to ask: can art interrupt or interrogate or perhaps invalidate this discourse? (Ruohonen 2003).

The recent popularity of impact studies in art education needs to be approached with some prudence. While such studies can be useful for articulating quality indicators and providing insight into effective practice, the application of impact research to drive advocacy is a dangerous approach. It opens the arts up to being a utility, only of relevance if its impact in terms of the latest educational trend can be proven.

Impact research should not be undertaken solely for the purposes of advocacy. Such an approach in arts education is likely to be self-defeating. Too much time and money spent on impact measurement detracts attention and draws valuable resources away from the improvement of art education for all children.

Furthermore, overemphasis on impact measurement situates the arts within a model of service rather than one of distinct value in its own right. To explain this point further, the arts have been all too ready to claim a host of positive benefits to fit into the latest educational bandwagon. For example, there was considerable hype over the possible connection between Mozart and IQ scores or in the role of the arts for intercultural understanding. This is not to say that the arts do not have a large number of diverse benefits, nor to say that the arts cannot be called into action in relation to a range of content or pedagogy. But importantly, arts education needs to maintain its inherent value and to argue the centrality of these values in the learning process.

The arts suffer from a tendency to lose focus by trying to be everything to everyone. To counter this, some arts educators have taken a very hard line against any talk of working across the curriculum. There is something safe and protective about remaining isolated within the art studio, music room or theatre space. Arts education has often adopted the mantle of "special," not allowing it to be fully integrated into the workings of a school. Similarly, some arts and cultural institutions have seen themselves to be "above" the needs of children and adolescents. The issue about permeability is not simply confined to education but can be more broadly applied to thinking about the arts and their place in contemporary society.

Conclusion

To conclude, I would like to stress the overwhelming importance of the arts in people's lives. Imagine living in a world where art played no role! In the hardest of regimes, the cruellest punishments are not physical but spiritual in nature—depriving people of their dreams, imagination, culture, history, rituals and things of beauty.

The arts provide us with the imagination to develop new ways of looking, living and making things. Art is the true "living treasure"—a triumph of mind over matter and an experience that makes us feel good and lifts our imaginations.

Teachers should not feel frightened of art or scared of their inadequacies. Enthusiasm is required, combined with an openness to see beauty and be moved. Cry over a piece of music, laugh at a clown, see yourself reflected in a portrait. In short be moved, and convey this gift onto the children you teach.

References

Aarons, A. 1991. A sociology of art...why does art look (or sound) like it does? *Artlink*, 11, 3: 36-39.

Anderson, T, E Eisner, and S Mc Rorie. 1998. A survey of graduate study in art education. *Studies in Art Education*, 40 1: 8-25.

Ashton, L. 1998. *I can't draw to save myself.* Australian Institute of Art Education, University of Wollongong.

Bamford, A. 2006. *The Wow Factor: Global research compendium on the impact of the arts in education*. Berlin:Waxmann Verlag.

Chia, J, J Matthews and P O'Shea. 1995. A window on an art classroom. *INSEA News*, 2, 1: 4-7.

Dissanayake, E. 1999. *Homo Aestheticus: Where art comes from and why.* Seattle: The Free Press.

Gadamer, H. 2004. *Truth and Method.* London: Continuum.

Hicks, J. 1993. Technology and aesthetic education: A critical synthesis. *Art Education*, 46, 6: 42-47.

Loveless, R. 1990. The electronic media conspiracy. *Leonardo*, 23, 2: 201-207.

Ranciére, J. 2004. *The Politics of Aesthetics.* London: Continuum.

Read, H. 1946. *The meaning of art.* London, Faber and Faber Ltd.

Ruohonen, R. 2003. *Creating Connections: Trends in Government Support and Artistic Creativity.* Report on the 2nd World Summit on the Arts and Culture, Singapore. International Federation of Arts Councils and Culture Agencies.

Scott, L. 1988. A Comparative study of personality, values, and background characteristics of artistically talented, academically talented and average 11th and 12th grade students. *Studies in Art Education*, 29, 3: 292-301.

Throsby, David, and Virginia Hollister. 2003. *Don't give up your day job: An economic study of professional artists in Australia.* Sydney:Australia Council.

Walker, S. 1996. Thinking strategies for interpreting artworks. *Studies in Art Education*, 37, 2: 80-91.

CHAPTER SEVEN

THE ART OF RISK IN AN AGE OF ANXIETY OR IN PRAISE OF THE LONG LUNCH

TONY MOORE

Anxiety can be good for art. I have fond memories of the culture of creative dissent triggered in the UK by Margaret Thatcher and her sociopathic Conservative government. British artists didn't whinge into their beer but rallied, critiqued and reaffirmed a lively human spirit. They mocked their rulers with post-punk agit-pop like The Clash and Elvis Costello. Movies and TV drama such as *Boys from the Black Stuff* and *My Beautiful Laundrette* interrogated class, ethnic identities and the secret state. Withering satire like *Yes Minister* and *Spitting Image* and the punk chaos of *The Young Ones* led a comedy renaissance. Call me a masochist but I preferred the angry, anarchic energy of the Thatcher-inspired Underground to the glitzy optimistic spin of Blair's Cool Britannia.

Closer to home periods of anxiety have evoked great cultural effervescence—think of the *Bulletin* writers and illustrators who found a rebellious yet humorous Australian voice during the great strikes and depression of the 1890s, the iconic art of the Angry Penguins modernists during the darkest days of World War II, or the emergence from the counter-culture of a new wave of local theatre and cinema in the late 1960s and early '70s—a period when Australia was mired in the Vietnam War, Cold War paranoia was rife and the conservative government seemed interminable. Provoked by "interesting times" these artists—drawn largely from subcultural margins—fashioned words, images and ideas that have helped develop a sense of what it is to be Australian.

Times of anxiety need not only produce nation-defining or left-of-centre clarion calls, but rather provoke interesting, radical art that shakes up pre-conceived ideas and aesthetics, art that is nihilistic, decadent, right wing, crazy—or just plain offensive—think Dada in World War I,

Weimar's Cabaret Voltaire, 1968 Situationists—or Punk. Liberated from the sepia-toned nostalgia with which we customarily cuddle yesterday's avant-gardes in the weekend newspaper supplements, these creative projects, at home and abroad, took guts to get up. They could be prickly, annoying, profligate, amateurish, pretentious, down right offensive and—risky. What all these artistic eruptions do share is a propensity on the part of artists and arts enablers to take risks. Without risk art atrophies and becomes a market category or nostalgia. For an example, look no further than David Williamson, who journeyed from edgy Carlton *enfant terrible* sensitive to the tragi-comedy of class mobility to bankable bard to the bourgeoisie indulging the colour by numbers obsessions of aging baby boomers. Or *Countdown,* a radical experiment in camp pop carnivalesque that died young with a full head of hair, only to be disinterred in the name of nostalgia to lumber around a stage "live" as a clumsy cadaver—called the *Countdown Spectacular*.

When John Howard was first elected in 1996 some of the artists I was interviewing for the Australina Broadcasting Corporation (ABC) documentary film *Bohemian Rhapsody* observed that a conservative government might be fuel for satirists and troublemakers. While a larrikin, subversive comedy undoubtedly thrived under the Coalition—think John Safran, *The Glass House* (bane of humourless conservative pollies) and *The Chaser* (our main security threat during the APEC clampdown)—I don't think the Australian arts community responded with the energy, iconoclasm and bravery we might have expected from our own age of anxiety. There are some notable exceptions—especially in writing and some theatre—but I don't get the feeling that Howard's version of Thatcherism triggered a popular art of dissent comparable to that of '80s Britain. Great edgy political books were written by the likes of David Marr, Margo Kingston and Ghassan Hage. Richard Bell's paintings pushed the Aboriginal resistance to new extremes and made white folks the object of derision. Scott Rankin's therapeutic Big hArt theatre projects, like the inspiring Knot@home, brought the actual victims of a dividing society, literally, to centre-stage. *Keating the Musical* reminded us of what had been and could be again, but this was the solace of nostalgia. But outside of comedy and newspaper cartoons "political art" has been a conversation among the converted, as opposed to the critical engagement with popular audiences that occurred, say, with the radical working class press in the 1890s and with Australian cinema in the 1970s.

Now that Labor has returned with a change mandate we are yet to see much of a response from the representatives of the arts community beyond the usual pleading for funding, and a new renaissance remains a hope rather than reality. Pop and rock music, in the past a site where bohemian flair and working class grunt fused to unsettle "the man," has succumbed to the manufactured idolatry of *Australian Idol.* TV drama and local cinema too often fail to tell stories people want to hear and are bogged down in the expensive aesthetic orthodoxies of the 1980s. While liberal America flashes its angry underbelly via a documentary renaissance that sets box office records, the genre has here disappeared into self-indulgent personal stories and info-tainment. Unlike the New Zealanders, Australians cannot do wondrous escapism like *The Lord of the Rings*, so trapped are we in the here-and-now certainties of realism.

I believe that Australian creative culture has lost its nerve, and as a result has entered a period of nervous quietism, where grey somnolent managers rusted onto thrones of influence set the national tone.

I know that there is no shortage of interesting grass roots creative people working within and outside cultural institutions straining at the leash to break rules, cause trouble and have fun with art; just look to the DIY larrikinism of the YouTube satire of the last election. Digital dada such as Hugh Aitkin's morphing of Kevin Rudd and his Labor comrades into Mao and the Red Guards easily outshone predictable mass media coverage. Sadly, too much new talent remains untapped while the same old stagers are paraded around for yet another lap of honour while they look back with nostalgia to the glory days of the last waves of the '60s, '70s and '80s. The broadsheets yawn with reminiscences about Heide and *Oz*, the Yellow House and the Pram Factory, *This Day Tonight* and the 1970s Australian Cinema Renaissance (I'm guilty of this myself), the very public slide nights of iconoclasts recycled as icons.

At the same time young people like Ryan Heath, pissed off Brussels-based twenty-something author of *F* Off, Its Our Turn Now,* chronicle how talented young Australians ambitious to make a mark in the creative industries have fled their homeland to take up opportunities in Europe, North America and Asia, where youth is no obstacle (2006, 143-154). According to Generation ExPat, Australia under Howard was starting to resemble a parochial country town where all the same old folks who've known each other for yonks run everything for themselves. When Heath wrote of such things in the newspapers prior to leaving Australia in 2003, Keith Austin bit back, declaring in the *Sydney Morning Herald*, "I'm sorry, Mr Heath, but you're going to have to prise these cold, dead fingers off my rung with your teeth, because I'm not stepping back to make way for you or any other generation." On Heath's return to Australia in 2006, Bob Ellis, rendered bereft of his customary eloquence at a Sydney lunch soiree I attended, simply threw a copy of Ryan's book at the young upstart.

I know how Ryan feels. A whole decade ago when I wrote an opinion piece in *The Australian* about our nation's cultural constipation I copped an editorial from *The Weekend Australian*, 23 February 1997, telling me to respect my elders and get back in my box because

> [i]n time the generation for whom Moore and others speak will become the dominant cultural force. By that time they will be no longer young. And they will be vastly more experienced. Which is how it should be.

So there. At the time I had several documentaries under my belt and was 35 years old (on the cusp of boomerdom but the ironic side of punk), which is middle aged if not an old fart in most Western cultures. What must be extremely frustrating for genuine young people now in their twenties, let alone 30s, is that the same baby boomer gatekeepers are still for the most part calling the shots in Australia's mainstream cultural institutions, while successive waves of youngsters wait outside the gate, coralled in the minstrelsy of comedy, pop music and game shows. But now Generations X and Y, with a few notable exceptions, are deserting a tepid arts mainstream for the fertile delta of alternative and new media.

The same editorial in *The Weekend Australian* consoled itself with "the younger generation's failure ... to swing the cultural highwire with anything like the panache of the boomers." The lesson of the transgressive art of the baby boomers is that younger people have to be prepared to turn their backs on how things were done then, and be original, offensive, rule breaking and unafraid of failing. But the problem is less a lack of ideas and creative nerve in the garrets and 'burbs where younger Australians make their YouTube videos, games and blogs, than a fear of new art, stories and aesthetics within the institutions whose business it is to enable art. We have to ask: why at a time calling for an art of risk, Australia's creative culture feels moribund?

The fear of risk has its roots in the managerial revolution of the early '90s, a cross-party blight, but it has been sharpened by growing authoritarianism in government and business at one end, and dependence of artists on shrinking state patronage at the other. And too many in the arts lobby revel in a childlike dependence on the public teat by bleating to Government for more funds. I agree that the Howard Government was parsimonious in its patronage of creative culture, but after a decade of austerity and the election of a culturally progressive yet "fiscally conservative" Labor government it's time we changed tack. Artists should reject the arts welfare model and be asking for the means to be entrepreneurial, because at its root being an entrepreneur means taking risks.

Artistic risk does not just happen, but is nurtured by cultural and institutional settings. We are living in risk-averse times, because the managerial culture that controls our public and private institutions demands accountability to accountants, value-for-investment measured by "indicators" and "outcomes" and top-down control of cultural assets. One only needs to look at our premier cultural institution, the Australian Broadcasting Corporation (ABC), to see a highly centralised, top-down management structure, where control of commissioning and editorial has in fact become more concentrated than it was a decade ago. Yet media technology for production and distribution are more suited than ever for heterogeneity, creative autonomy, narrowcasting and interactivity. Two examples from TV will suffice. The internet allows meaningful audience involvement in programming, from ideas gathering and commissioning to themes, stories and character development. Online audience participation is increasingly encouraged by commercial TV and the ABC, but their programming persist in ignoring audience diversity of interests, appealing instead to lowest common denominator ratings and outmoded executive hunches to shape their content. In the under-performing area of TV drama we hear constantly about its high cost, but if we dare to ditch the glossy aesthetics and production values of the '80s and utilise digital cameras and guerrilla crews, like Sydney's TropFest Short Film Festival entrants or ABC comedy, so much more would be possible on available funds. Sadly playing it safe is what gets rewarded in our managerial age.

The worst aspects of this managerial ascendancy in cultural institutions, embraced by both Labor and Coalition governments at federal and state level since the early '90s are:

- A tendency to view senior management and in some cases government as the principal stakeholder rather than audiences and cultural producers.

- An intolerance of the mistakes and failures that a risk-taking culture would applaud as a necessary by-product of experimentation.

- Underestimation of intangible, unquantifiable social formations that greatly enhance creativity—everything from student life and bohemia to laziness and long lunches.

- Adherence to commercial indicators of success that simplistically divide diverse audiences into masses, elites and core audiences.

- A passive model of arts consumption rather than encouraging popular production and interactive appreciation. The success of YouTube demonstrates that younger audiences enthusiastically make their own media content, yet most traditional cultural institutions that work on the drip-feed model lament that the young have deserted them.

- The entrenchment of an aged cohort as cultural gatekeepers who snared all the good gigs in the '70s and who are closed to new talent experiences, ideas and aesthetics.

- The circulation of the same group of managers from one cultural institution to another, preventing real change. For example, in the public media sphere a small pool swish from the ABC to Film Australia and from SBS to the NSW Film and Television Office. This game of musical chairs gives the appearance of change, but no real change.

- Too many of the public spaces of Australian arts, such as art galleries, drama theatres and orchestra concert halls remain rarefied, church-like, bourgeois, in a word "posh," thus alienating artists from their fellows.

- A failure to comes to terms with Australia's class differences and ethnic diversity in subtle, nuanced ways. As a result of the background of most senior managers our media is anachronistically Anglo-Celtic.

- A colonial mindset written deep into the DNA of many cultural institutions that marginalises new, local stories and Australian traditions at the expense of Northern imports, especially from the UK and US.

- Lack of education in Australian cultural history which limits managers' vistas to their own biographies so that we are condemned to a creative *Groundhog Day* where we are perpetually reminded about the time of youth of the currently middle-aged but kept in ignorance of what came before and after. Thus, a recent ABC stab at history, *How the Hell Did We Get Here?*, once again returns us to the steps of the old Parliament House in 1975 on the "day the renaissance died," without ever wondering what we owe the nineteenth century.

- An inability to appreciate that in a diverse society—criss-crossed by age, class, regions and ethnicity and where people have different cultural literacies and interests—art will necessarily offend some people, leave some indifferent, and deeply engage others.

While both Labor and Coalition federal governments have run with managerialism, state Labor Governments have at least supported the arts as an engine of economic growth and Australian dreaming. The Howard Government was not a great friend to the arts, a flaw reflected in actions such as the funding squeeze on the ABC, the introduction of a goods and services tax on books, censorship and outright banning of films, and the continuous disparagement of arts and intellectual workers. That Government had a philistine tinge that ultimately ran contrary to its pride as an enabler of economic growth. John Howard used to say he favoured entrepreneurs. Yet his government's neglect of art and hostility to artists, obsession with an homogeneous cultural mainstream, and social conservatism hostile to transgression were the greatest obstacles to an entrepreneurial spirit in the New Economy, and held back Australian capitalism from realising its potential in the post-industrial age.

The election of the fresh-faced Rudd Labor Government pledged to modernise the economy and invest in "human capital" gives us pause to hope that cultural enterprise will be encouraged as a driver of the economy and national dreaming. Now that Kevin Rudd has asked them, perhaps young, Generation ExPats will return as they did in the '70s on Whitlam's ascension? Would they get a good gig if they did? The ever present danger is that Labor's long embrace of managerialism, and the bureaucratic background of so many of its frontbench, will cruel the government's capacity to clean out the managers who cripple risk-taking in the arts. Luckily the relevant Minister, Peter Garrett, cut his teeth as a rock 'n' roll rebel in the post-punk Ozrock renaissance of the 1980s. Rather than a bureaucrat, he ran a successful music enterprise called Midnight Oil and knows the value of risk-taking. Not surprisingly then, the Labor Arts policy sees the arts not as a charity case but as a generator of wealth and contains measures to assist "cultural entrepreneurs" accumulate cultural capital (Garrett 2007). This approach bodes well.

As Richard Florida (2003) argues, encouraging the arts and cultural enterprise is the engine of economic growth in the new economy. If Australia is to sustain its standard of living and employ its people it must become an exporter of value-added products. Like Ireland, Australia is an English speaking "rock and roll" country ideally placed to build on its successful export of higher education by also exporting television, film, music and online content to the world. Modern Western economies are diverse, and increasingly based on information—not just housing property, primary produce and minerals—yet the Howard Government played to the past. Their restriction of digital television to protect the existing media oligopolies signalled their "old economy" mindset and fear of risk. Australia could have diverse narrowcasting media distribution channels in both the commercial and public sectors, fed by a huge cottage industry. But instead we had protection of media mates, scared of competition. Australia's media laws are the antithesis of free market reform, designed to shore up the profitability of the existing cartels, especially the dinosaur, free-to-air TV networks that are losing viewers. Let's hope the new government has the guts to bust open this cosy club.

The late French sociologist Pierre Bourdieu liked to look at artists as cultural entrepreneurs, who not only compete as individuals, movements and generations, but also amass and deploy capital (1993, 68 & 195). But this capital is not money, as industrial capitalists wield, but "cultural capital"—the mix of knowledge, skills, contacts, style, personality traits, ideas, education and marketing savvy that enhance the raw talent gleaned from genetics and upbringing to make an artist not just good, but able to work and get noticed and work again. Cultural capital is embroiled in media and reputation and may become notoriety, fame and celebrity. Institutional arts managers and governments understand quantifiable enhancers of cultural capital—like education and training, or galleries, theatre companies and TV stations—but they miss those equally important banks of cultural capital that cannot be measured or controlled.

Some of the most important of these are extra-curricula tertiary student life among the young, and the formation of bohemian groups and avant-garde movements.

Australia has a rich bohemian tradition stretching back past the postmodern subcultures of the 1980s and '90s, Punk and hippy counter-cultures of the '70s and '60s, the libertarian Sydney Push and Melbourne's Carlton, the Angry Penguins' surrealism and the social realists of the 1940s, Norman Lindsay's vitalist circle, through the Heidelberg Impressionists and *Bulletin* bards of the late nineteenth century, to Marcus Clarke's dandyish bohemian clubs in the 1860s and '70s.

Bohemia has great value for artists, not least as a crucible where class and ethnic differences are hybridised, and as stimulating space to nurture transgressive art. Bohemia and especially the more self-conscious avant-gardes are great ways for the unknown young artist to make contacts and generate a buzz in the city. Bohemias are escalators of social mobility, merging at one end with the *demimonde* and deviant and at the other with the bourgeois customer and political class. The common bohemian pastimes of intense conversation, showing off, dining, drinking and drugging, and observation and experience of the urban spectacle enhance networks and creativity. Being a bohemian is a licence to take risks. Bohemia gives the cultural producer the touch of glamour and danger that the bourgeois consumer craves, if only for one nice night's entertainment.

But bohemia is only one of these intangible engines of cultural capital. Workplaces can encourage a bohemian-like environment. The *Bulletin* and *Smith's Weekly* encouraged camaraderie and community. The ABC used to be the home of the long lunch and crazy experiments like Triple J and *Beat Box* in the 1970s and '80s. Beyond work we should realise the influence on creativity of humanities education, religion and transcendent experiences, travel, being multi-lingual, ethnic cosmopolitanism, madness, and Mirka Mora's favourite—laziness.[2]

While such formations cannot be controlled, government and managerial action can enable some of them, and they can certainly be damaged or destroyed to the detriment of creativity in the community. I don't think HECS-HELP debts incurred by tertiary students, the ending of compulsory student contributions to campus unions, and high rents that compel students to work a number of part-time jobs and live with their parents have encouraged bohemia on campus. Late twentieth-century managerial models—beloved of the Australia Council and government film funders—that seek to account for and control all assets actually destroy cultural capital.

For me the role of the state and public intervention is to encourage diversity of channels of production and distribution, and ensure a fair democratic spread of cultural capital. Further, the state should encourage, within public and private institutions, those intangible networks that build cultural capital and audiences.

The Labor Government has promised to establish an "Innovation Centre" to "incubate a cluster of small and medium-sized Aussie arts companies and creative businesses," and to help them access the latest creative technology. Governments can also build and distribute cultural capital through the provision of life-long, high quality public education and I hope Labor's much vaunted "Education Revolution" will encourage extracurricular activity in schools, TAFEs and universities. The two to three hours of school-based aftercare before working parents pick up their children could be a site for extra-curricula instruction in music, creative writing, digital arts, school newspapers or theatre. Many private schools already use aftercare to lift their kids' creative skills in this way. At the other end of the education experience, restricting student social life by kneecapping university student unions, for instance, is going to have a negative impact on creativity. *Aunty Jack* and *The Chaser* both came from student revues.

Equally important is the nurturing of talent in public and private media spaces that should be a diverse cottage industry rather than twentieth-century style silos. Simply throwing more money at the arts is not the answer if it goes to the same old same old, to be gobbled up by bureaucracy or used to indulge the tastes of a tiny percentage of the population. Public policy should stimulate networking and venture capital for hothouse projects rather than be poured into monolithic institutions that squander resources on consultants and management hierarchy. Why not provide low interest government loans for artists and cultural enterprises that mimic the micro-credit schemes enabling self-sufficiency in the developing world? Or government funds could be directed to a much larger number of start-up projects, with extra public money kicking in on a dollar-for-dollar basis as private investors come on board (a version of which will occur with the new Australian film tax concession to be administered by the new Screen Australia). At present too much public money is granted on the basis of the arts bodies mimicking corporate management practice and squandered in institutional red tape and compliance overheads, rather than rewarding the essence of business—a propensity to take risks. Labor needs to re-examine the criteria by which funding bodies such as Screen Australia give projects the nod, and cut out the deadwood from the existing cultural boards, ensuring that the diverse tastes of younger generations are represented.

I have no doubt that a centralised top-down control structure and elevation of the bean counters has cruelled the ABC's creativity and nerve to take risks. Increased funding will be wasted on more bureaucracy unless managerial hierarchies are flattened and creative decision-making devolved. Let's return to a devolved model where department heads and executive producers commission for dedicated timeslots. What's to lose if an ambitious risk doesn't work?—it's only TV, for god's sake, not surgery!

It is admirable that the ABC is embracing a multi-channelling future despite government parsimony—a development that has the potential to enhance choice, audience participation and creativity. So far the digital channels are still behaving like broadcast channels, and recycling *The Bill* and *At the Movies* from ABC 1. But there is movement at the station, with ABC 2 giving a prime time slot to the anarchic *Double the Fist* and other edgy shows. Having multi-channels like the BBC will potentially allow "Aunty" to cater for the different niche tastes of her numerous nephews and nieces without offending each others sense of quality, beginning with a kids/youth channel. However these will require additional resources if they are to be fair dinkum and Rudd the fiscal conservative was careful not to promise funds for a Triple J of TV as Lindsay Tanner did in 2004 (2004, 2).

ABC online and new digital stations have the potential to enhance audience participation in our ABC. While many an older, rusted-on ABC viewer might have sneered at *Big Brother*, this one-time Channel 10 hit demonstrated that the younger audience wanted to participate in the narrative direction of a TV show. Building on its own pioneering efforts in audience generated content like *Beat Box* and *Race Around the World*, the ABC should allow audiences to shape content still further via online, commenting on pilots or even offering up their own pilots for comment,

"YouTube-style," as children already do on the Rollermache site. Beloved participatory children's radio series, *The Argonauts*, could sail again, only this time by fibre optics and a set-top box. This is a democratic and creative trend collapsing the barriers between consumers and producers, and it should be a priority for resourcing by the Rudd Government that pledged in its 2007 Platform to ensure that the ABC be "able to exploit the potential of new technology" (2007, Ch.16, Point 57).

A new cultural renaissance will not be built by bureaucrats subsidising elitism or "picking winners" to passively entertain us, but by entrepreneurs and public institutions bold enough to harness the diverse creative energy in the community, from suburban garages to inner city garrets.

Social democrats in state and federal government and the public arts bodies need to avoid Keating-era notions that divide artists off from the rest of us as a special caste and ask, "how can we spread creativity around?". I am critical of the persistence of the romantic idea of the "artist hero" in Australian arts discourse and of the disconnection of Australian arts from contemporary working class life. We need to counter the dichotomy of "artist" and "philistine masses" with an appreciation of the creativity of popular cultural forms, craft skills and the role of audiences in creating value for art.

In place of the passive arts consumer I want to emphasise the idea of all people as creative producers—a replay of the William Morris idea of do-it-yourself arts and crafts for the Digital Age. Youth subcultures, like culture jammers, hippies, punks and community hobbyist organisations, have sought to empower people to make their own art, from folk and rock 'n' roll bands, to "indie" record labels, to amateur dramatics, YouTube videos and blogs. Cheap, interactive media technologies have the potential to enfranchise a do-it-yourself creative culture in the suburbs.

I don't mean a *Wayne's World* of amateurism, but genuine, democratic talent scouting; a 21st century version of the approach the *Bulletin* of the 1890s, that scoured the bush and the back lanes looking for poets and artists and discovered the likes of Henry Lawson, Banjo Patterson and Norman Lindsay. Academic Sylvia Lawson has shown that it was an interactive dialogue with the *Bulletin*'s readership in the cities and the bush that created a passionate community around the project in the late nineteenth century, and produced some of the magazine's best writers. (1987, ix-xii) Its founding editor and co-owner, J. F. Archibald, argued that everyone has at least one good yarn, poem or drawing in them and put his money where his mouth was, inviting contributions from ordinary folk and paying published contributors. The popular creativity the *Bullie* unleashed in those years recalled the blending of high and popular culture that occurred in Elizabethan theatre. Tapping into shearing sheds, docks and pubs unearthed house-painter-cum-poet Henry Lawson, and, perhaps our greatest novel, Joseph Furphy's *Such is Life*! The *Bulletin* of the 1880s and early '90s could be politically radical in its championing of the common man and causes like republicanism, unions and democratic reform (just as it could be sexist and racially prejudiced). But it is in the magazine's carnivalesque disruption of expectation and its topsy-turvy mockery of authority, blending of genres, use of cartoons and humour, self-conscious parody of the media itself and blurring of readership and contributors that the *Bulletin* was most subversive. The best Australian culture erupts when media enfranchise passionate communities and become clearing houses for new ideas and styles, as happened again with early 70s cinema, *Nation Review*, Double Jay, inner city post-Punk pop and even *Countdown*.

Today's managers need to consider the relationship of content producers and distributors to their audiences. Australian arts discourse, enamoured of the artist hero or "star" does not pay audiences sufficient regard. Audiences not only add value by the way they make sense of art, but are also a source of ideas and aesthetic innovation—the suburbs are much weirder, more interesting and diverse than gatekeepers and programmers obsessed with a mythical mainstream could ever imagine. One of the drivers in the young's passion for digital mobiles is that they can use them to contribute to their own made cultures. Out in the 'burbs young people are assembling multi-identities from the material surrounding them as they grow up—the immigration experience, family, religion and politics, sexual options, global TV, the internet, blogs, music culture and neighbourhood. Cultural diversity—ethnic, lifestyle, regional, religious—is the great narrative of contemporary Australian life, but apart from *Pizza*'s swaggering ethno-larrikin comedy and Chris Lilley's timely satire of triumphalist Aussie values and PC tokenism, it makes barely a ripple in our media or art. The clash and fusion of cultures in the suburbs is a new wellspring of hybrid art; just as Cubism, jazz and rock rolled out of the cities in the last century.

How can Australia ensure that young people from working class and other marginalised backgrounds have the opportunity to develop their creative potential and contribute to the new cultural economy? This is where Labor's Education Revolution needs to move beyond computers and trades. It's worth considering the experience of the British Art School that helped translate generations of working and middle class kids, who did not fit into scholarly or technical education, into commercial designers, pop musicians, film makers, fashion gurus and artists. Think *School of Rock* for children from blue-collar families. Skilled up for creative entrepreneurship and expression, working class sensibilities can change a national culture. Graduates of the British Art Schools include John Lennon, Paul McCartney, Pete Townsend, Malcolm McClaren, and Vivien Westwood. East Sydney Tech, Sydney's closest equivalent, gave us Martin Sharp, Jenny Kee, kooky cartoonist Reg Lynch, Mental As Anything, and, via Reg Mombassa, the genesis of Mambo. Swinburne in Melbourne has churned out some of our most innovative film makers and commercial artists. Reformed public schools and TAFE can be the site for teaching cultural literacies and skills, and building a democratic creative culture in Australia.

At present far too many of our film makers, journalists and artists come from well-off, upper middle class backgrounds via elite institutions like universities, the Australian Film Television and Radio School and our National Institute for Dramatic Art. The trick is to keep these arts finishing schools and Australia Council grants at the apex of the pyramid, but to build on a base of democratic creativity fostered through the public education system.

Veteran art historian Bernard Smith asked this of a *Blue Poles*-smitten Whitlam Government back in the 1970s (1988, 7) and it is even more relevant today when cultural industries are becoming economic powerhouses. The great Australian cricketers and swimmers arose from a vibrant participation of all kids in school and local team sport. Smith reckoned everyone is an artist, just as we can all play sport, and recommended that government intervention go into energising the arts instruction in state schools, including fine art, creative writing, music and so on. Most kids love to draw, paint, sing and tell jokes, but their natural talent and inclination gets knocked out of them as they move through high school. "If your object is excellence," Smith observed, "you begin with a broad base: elitism fails because it insists upon a narrow base; its apex fails to reach the limits of the possible." (1988, 7)

References

Austin, Keith. 2002. The Heckler: Life's tough, Xers, but you'll get a hand if it's deserved. *Sydney Morning Herald*, 1 February.

Australian Labor Party. 2007. ALP National Platform and Constitution. 44th ALP National Conference, 27-29 April.

Bohemian Rhapsody: Rebels of Australian culture, First broadcast December 1997 by ABC Television. Directed and written by Tony Moore.

Bourdieu, Pierre. 1993. *The Field of Cultural Production. Essays on Art and Literature.* Cambridge: Polity Press.

Florida, Richard. 2003. *The Rise of the Creative Class: and how it's transforming work, leisure, community and everyday life.* North Melbourne: Pluto Press.

Garrett, Peter. 2007. *New Directions in the Arts: Supporting a vibrant and diverse Australian arts sector*. Australian Labor Party.

Heath, Ryan. 2006. *Please Just F* Off, its our turn now.* North Melbourne: Pluto Press.

Kirkpatrick, Peter. 1992. *Sea Coast of Bohemia, Literary Life in Sydney's Roaring Twenties*. St Lucia: University of Queensland Press.

Lawson, Sylvia. 1987. *The Archibald paradox, a strange case of authorship*. Ringwood: Penguin

Moore, Tony. 2000. Australia's Bohemian Tradition. In *The Australian Legend and its Discontents,* ed. Richard Nile. St Lucia: University of Queensland Press.

—. 2001. Unchaining Aunty. In *For The People*, *Labor Essays 2001,* ed. Dennis Glover and Glenn Patmore Annandale: Pluto Press.

—. 2004. Australia's Bohemian Carnival. *History Australia, 2: 1: pp. 11.1 to 11.17.*

Smith, Bernard. 1988. Notes on Elitism and the Arts. Published in *The death of the artist as hero: essays in history and culture.* Melbourne: Oxford University Press.

Tanner, Lindsay, and Kate Lundy. 2004. A Better, Stronger, Independent ABC—Labor's Policy for the ABC. Australian Labor Party Policy Document.

Wilson, Elizabeth. 2000. *Bohemians: The Glamorous Outlaws.* London: I.B. Taurus.

Notes

[1] Disclaimer: Tony Moore commissioned Ryan Heath's *Please Just F* Off* in his capacity as Commisioning Editor of Pluto Press Australia
[2] In an interview for the ABC documentary *Bohemian Rhapsody*, Mora explained to the author the creative muse of relaxing and just doing nothing.

PART III:

REGIONAL DEVELOPMENT

CHAPTER EIGHT

ECONOMIC DEVELOPMENT THROUGH THE ARTS IN RURAL MONTANA

JOHN C. BARSNESS

Introduction

In Montana economic development through the arts is going everywhere and nowhere at once: it is a process similar to herding cats. There is no central coordination and the mix of advocates and practitioners includes state government agencies, private entrepreneurs, non-profit and for-profit organisations, and individual artists. Their efforts take off in a multitude of directions, sometimes work against one another and, like cats, sometimes ignore whom they please. Toss into that mix ten Native American tribes and seven reservations, all of which are not only sovereign entities but also have strong connections to state and federal governments, and things become truly interesting: one might describe it as pure democracy in action.

For reference and perspective (see Fig. 8-1), Montana is located in the northwestern United States, shares its northern border with three Canadian provinces, spans the entire Mountain Time Zone and is geographically isolated from major population centres. Montana's total population of just over 900,000—approximately that of the city of Tucson, Arizona—is spread over a geographic area nearly that of Japan (145,552 square miles) or approximately one half the land area of Australia's New South Wales. Population density is 6.2 people per square mile, and a *Newsweek* Magazine cover article in October 1989, referred to Montana as "America's Outback".

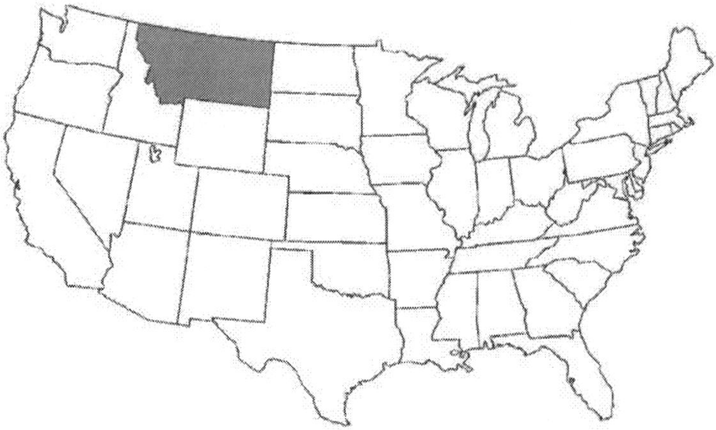

Fig. 8-1 Location of Montana in the USA

It is a vast, geographically isolated, sparsely populated state with many natural resources. The western third is mountainous and contains Glacier National Park in the north, is a major gateway to Yellowstone National Park in the south, and between has several designated wilderness areas. The eastern two thirds contain mainly high plains and open spaces that truly can be described as "Big Sky Country." These natural attractions also provide the basis for the state tourism agency's out-of-state promotional campaign. Unfortunately, as with many remote communities, images of unpopulated areas imply nobody lives here and therefore we have no culture—when in fact quite the opposite is true.

Montana has a significant Indigenous population and important manufacturing, agriculture and service economies, as well as experience with "boom and bust" extractive industries.

Australians are familiar with the notion of an export-driven "colonial economy", and through most of its relatively short history Montana has had exactly that relationship with the rest of the United States. In the late 1940s, because of a decline in metals mining, Montana fell from fourth in per capita income to near the bottom among US states today. The resulting lack of economic opportunity created an outflow of talented youth who departed to find employment in other states and rarely returned. In his book *Collapse*, Jared Diamond describes how 70 per cent of young people in Ravalli County, Montana, left to seek employment elsewhere (2003,

63). Moreover, six straight years of drought severely hampered the State's agricultural base prior to 2005, with lingering effects in subsequent years.

But the "Digital Revolution" has begun to reverse this process, as technology allows people to live wherever they please and telecommute.[1] A decade ago in my home city of Bozeman there were four high-tech businesses; today, there are more than ninety. Quality of life is one of the main factors drawing new businesses to the State and growth in jobs, with related construction and development, helps keep young people from leaving. The downside is that locally home prices more than tripled in the past decade and some parts of the State are becoming affordable only to the wealthy.

Throughout its brief history, Montana's ranches had to be large in order to survive economically. At minimum a ranch had to be 2,500 acres in size or larger, due to the combination of soils, short growing season, and semi-arid conditions. Whether growing wheat, or raising cattle, financial success in agriculture depended upon large-scale operation. With economic pressures from global competition depressing market prices, the impact of drought (and climate change), taxation and increased demand for housing, the landscape is changing.

Montana's mountain scenery is stunning and has proved to be a major attraction for eco-tourism as well as a reason for the recent development of "exurban sprawl" near several western cities. The appeal of a twenty-acre ranchette that does not require a subdivision zoning review means new homes sprout as fast as mushrooms in former agricultural land. The dilemma facing financially strapped ranchers is a decision whether to maintain their family ranching heritage or to accept an offer from a land developer and become an instant multi-millionaire. This process of "gentrification" is changing the population mix as wealthy people from out-of-state build large vacation homes near some western cities.

For example, in the western third, several idyllic valleys that through most of the twentieth century had been devoted nearly completely to agriculture are filling with housing developments. In addition, in part due to a federal program of consolidating checkerboard land ownership left over from nineteenth century transcontinental railroad expansion, forest lands have been swapped or sold and opened for development in some areas. The result is increased construction of recreation homes in former scenic timber land. In one example, a private developer created a gated community and ski hill bordering a wilderness area that requires a minimum net worth of $3 million USD just to apply for membership.

So, despite promotional photographs and postcards depicting unpopulated scenic vistas, Montana is experiencing a growth problem with population expansion at its root. It is a complicated issue, however, one of the contributing factors is the economic success of people living elsewhere realising the American Dream of owning a home in the country with plenty of space around it. In other words, people may be loving the state to death.

Arts and Development Approaches

Creative industry in Montana includes a wide range of participants: 2,088 businesses employing at least 8,424 people were registered with Dun & Bradstreet in 2005. And these figures do not include those visual artists, sculptors, crafts people, photographers, musicians, dancers, other performing artists, theatre companies, architects, museums, galleries, film, radio, television, design and publishing, schools, creative service businesses and non-profit organisations not registered with Dun & Bradstreet.

Two development approaches that arts organisations, including Montana Arts, and local practitioners have been engaged with extensively in the region are "creative clusters" and cultural tourism. In these arenas, much has been done; but far more remains to be accomplished.

The Creative Clusters "Experiment"

The New England Council's 2003 Creative Economy Report usefully defines a "creative cluster" as a group of organisations and professional entities—including non-profit institutions, commercial businesses and individual artists—producing goods and services based in cultural enterprise, the fine or applied arts.

In 2002, the Montana Department of Commerce commissioned Dr Stuart Rosenfeld to examine Montana's economy and make recommendations for new development initiatives. His report to the Department of Commerce identified six business clusters including two—cultural tourism and creative clusters—where the arts figured prominently (Rosenfeld 2003). The Department subsequently decided to pursue two of these, merged into one, creative and wood products industries.

In consequence, there are two separate, but related, creative cluster projects which developed and were connected through a single citizen appointee working with both.

The first is a state-level task force involving the Montana Arts Council working with the Montana Department of Commerce as a partner in growing professional development opportunities for Montana artists, through their "Made in Montana" grant assistance program and other partnerships in development.

The second is a loosely knit, grassroots effort based in Missoula. At start up, this project held three meetings over a four-month period and invited representatives from creative businesses, government, education, and non-profit fields and approximately 100 people participated.

These Missoula meetings featured introductions with a brief description of current work, then an in-depth look at a single creative project. At the second meeting, local musician and software developer, Ron Newman, introduced his concept of the "IdeaTree" for "modelling complex relationships, from business processes to ecology", and for "sharing of data across institutional boundaries" (Newman 2002). The local arts agency used this program to plan creative cluster meetings.

The grassroots meeting format not only provided a good excuse to meet (beer and snacks provided!), it was also a forum for people with similar interests working in creative endeavours—but in isolation— to meet and exchange information. There were useful connections made outside the creative field. State and local government representatives attended the three meetings and saw first-hand how many people were participating. People working in education developed contacts with businesses that could provide internships for their students in creative fields. Software developers and other high-tech business people met and swapped stories as well as possible areas of collaboration. A non-profit arts organisation found a source for a high-definition computer and monitor for use in an international visual art exhibition. A Yellowstone Public Radio reporter met several people who would become subjects for feature programs.

However, the creative cluster development program stalled when State

Government control changed parties in 2005 and the bipartisan effort the previous Governor promoted did not enjoy the same support amongst the incoming administration from the opposite party. So, after an initial productive flurry of activity, a rigorously researched and popularly supported plan for development and promotion of Montana's creative and wood products industries was abandoned.

Instead, the new administration passed and promoted its Big Sky on the Big Screen Act, which gives tax incentives and rebates to out-of-state motion picture companies—rather than support growth of 8,000 to 10,000 long-term tax-paying jobs in-state. This choice—with its focus on short term location "windfalls" for a handful of communities—appears to continue Montana's infatuation with the "boom and bust" economic strategies of the past.

The lessons learned: our State Government moves slowly and deliberately but may eventually commit resources to foster creative business development growth. Moreover, since it has been said that there is no "Montana economy"—only a collection of small regional economies existing within the same state's boundaries—grassroots connections, as happened in Missoula, will continue to be the key element in developing these local economies.

Trialing "Cultural Corridors" for Tourism and Audience Development

According to a Travel Industry Association of America theory presented at the 1998 Americans for the Arts Cultural Tourism conference in Atlanta I attended, there are only two kinds of tourists: environmental and cultural. Moreover, while the concept of cultural tourism means different things to different people, a local, working definition is: the best experiences and opportunities Montana communities can offer that give visitors something they cannot get at home.

Montana's main tourist destinations are Yellowstone National Park and Glacier National Park. According to Norma Nickerson's 2005 research for the Institute for Tourism and Recreation Research almost ten million people visit our State every year. The study also noted a miniscule 0.6% average growth in non-resident visitors over the 2000-2004 period—far below national trends—and concluded that potential visitors were lost to those states with higher advertising budgets (Nickerson 2005).

In a previous ITRR study from 1998, out-of-state residents who did not plan to visit Montana were asked for reasons why. Their overwhelming responses, in order, were a perceived lack of cultural resources and unique shopping opportunities (Parrish 1998).

Of non-resident tourists surveyed in 2005, 20% visited museums, art exhibits, plus performing arts events and 19 % historic sites—well ahead of gambling at 6% and golfing at 4% (Nickerson and Oschell 2005). Given the large amount of recent community and private investment in casinos and golf courses, and the relatively small interest from tourists, the potential is for more positive impacts stemming from the promotion of Montana's existing cultural attractions and events.

According to the Travel Industry Association of America's 2003 research, the average tourist in the US spends $457 per day and a cultural tourist spends around $623 per day. However, according to a 2008 report from the University of Montana, the average non-resident visitor spent only $153 per day in Montana (Rademaker 2008, 18). This means that they are currently purchasing fuel, meals, lodging, and little else.

The research points to an opportunity for Montana to promote its overlooked cultural assets to the lucrative cultural tourism market. Currently, Montana is positioned in out-of-state markets as a destination almost exclusively for environmental tourists. However, the cultural tourist potentially offers a greater return on investment.

Until 2005, when fiscal crises in other states resulted in deep public funding cuts, Montana traditionally sat dead last in government support for the arts (56 of 56 states and territories), had no Fortune 500 headquarters, and only two significant private foundations that gave insignificant amounts to the arts. Therefore, audiences were seen as the only source of revenue available with potential for expansion.

When, in 1998, Montana Arts received a National Endowment for the Arts Planning and Stabilization grant to assist with specific training in audience development, representatives from 47 cultural organisations met and decided the definition of "audience" must expand to include people in other communities, counties, and states. The group then reached three conclusions:

- They would work individually to expand their audiences.

- They would work with one another and create logical alliances. For example, organisations in the small rural towns of Choteau and Fort Benton (approximate population 1,700 each) paired with the City of Great Falls (population 50,000) with the goal to help develop each other's audience through communication and planning, cooperative scheduling of events, and sharing mailing lists.
- Montana Arts staff would work on the long-range goal of increasing out-of-state audiences for the arts through cultural tourism development.

A "cultural corridor" is a marketing tool resembling the more familiar concept of a "heritage corridor", but with a broader focus that includes mapping visual and performing arts. Many US states have used heritage corridors successfully to promote local historical attractions. A working definition of a cultural corridor is any group of cultural attractions in relative proximity to a transportation route—which may be a few city blocks to several hundred miles in length.

The idea of a cultural corridor is not new, the concept was made popular in the US in the early 1990s. Although there are many successful examples, perhaps the best known is Becky Anderson's "Hand Made in America" craft corridor in North Carolina.[2]

During 2002, Becky Anderson worked on the development of the "Hands of Harvest"[3] north central Montana loop tours: one of our State's three emerging cultural corridors. With the other two, Missoula Cultural Council is promoting a set of tours in western Montana and Montana Arts have developed a Highway 200 Cultural Corridor, linking eastern and central Montana. But the tours have to be more than simply lines on a map: a wealth of detail, both in text and images, of what is available and how to find everything listed must be included in all public materials.

The Highway 200 Cultural Corridor stretches two thirds of the way across the state and includes towns of Sidney, on the eastern border, Glendive, Circle, Jordan, Winnett, Grassrange, Lewistown, Hobson and Belt. During 2002–3, Montana Arts staff created a "cultural inventory" as a basis for the corridor and a website that conceptually and graphically links these communities. This site will provide one-stop shopping for tourists planning vacations that include Montana and who want to include cultural activities on their way through the state.

However, the question of "critical mass" remains—are there sufficient cultural resources available to the travelling public to make this route a worthwhile detour? In 2008 the site remains short of necessary funding to proceed and may include additional natural resource sites to expand its appeal to both the cultural and eco-tourist.

More lessons learned

In general, there exist some common threads successful cultural tourism projects share in the US. These include:

- A concentration of existing cultural resources and organised experiences.
- Attractions within an easy (one-day), round-trip driving distance from a city with population greater than one million, or exist in a region with high population density that has year-round access to the attraction.
- Concentrated and available resources from corporations and private foundations committed to financing cultural tourism development.
- Sufficient public federal, state, and/or local tax dollars are committed to sustain staff and marketing.

In Montana we have almost exactly the opposite conditions:

- Montana's cultural resources are widely scattered and there are many unadvertised cultural experiences available.
- The out-of-state audience is diffuse and geographically distant; there is no city with a population greater than one million within a one-day round trip drive of any of Montana's cultural attractions.
- There is next to nothing in the way of corporate or private foundation resources that could be focused on cultural tourism development.
- State "bed tax" dollars primarily support advertising to attract environmental tourists and the state tourism agency has no staff dedicated to cultural tourism.

Part of our research for the Highway 200 Cultural Corridor consulted the people who worked on the development of the 143-mile-long Pennsylvania Schuylkill Heritage Corridor. Their annual budget was $3 million (USD), or $20,000 per mile, while Montana Arts had just $3,000 for 370 miles—$8 per mile.

A CBS News reporter once spoke with me about "the power of the naïve," wherein one succeeds by not realising the task is impossible. In many respects our Montana projects must work according to this principle.

One of the most difficult realities for an outsider to grapple with is just how limited the resources are in small rural communities. Many local museums are staffed entirely by volunteers and able to open only on weekends during the peak summer season and not able to open during the rest of the year. Irregular hours do not attract or please tourists. It is a classic Catch-22, the Corridor aims to bring more money into a community but the community does not have the money to support the cultural attractions being promoted; and if tourists do not stop they do not spend their money.

The "Hands of Harvest" organisation, assisted by Travel Montana, created seven loop tours using the City of Great Falls as a travel hub and published a guide book, which is now in its second printing. On these tours one can see out-of-the-way attractions as well as visit artists' studios. The organisation learned the value of having a well-organised, motivated group of volunteers. The area's economy is heavily dependent on agriculture—which is subject to extended periods of drought and fluctuating world market prices for wheat. With depressed agricultural prices and reduced harvest, in many cases, an arts-related home business can bring in extra money to help families keep their farm or ranch. It has made sense for proprietors of such businesses to help make the cultural corridor a success

There is no one correct way to create a cultural corridor. In all three Montana cultural corridor projects and to varying degrees, it was a single person who acted as the "spark plug" to keep momentum going. The most successful had state tax support and a committed group of volunteers; in the other two both the state funding and volunteer resources were missing. The latter projects also have the most unfinished work remaining.

In those projects that include Indian Reservations or native populations, to achieve success, participation in a cultural corridor must truly "be their idea," and local public planning meetings are critical. The idea for a corridor can come from outside but the shape it takes must be decided locally; and native people must decide for themselves exactly how much of their culture they wish to share with outsiders.

Whatever is designed must be inclusive. Without the resources that urban areas take for granted, our corridors must package and appropriately market the range of cultural products available: high quality artist studios, art galleries, art and historical museums, performing arts events, humanities events and resources, art history, historic sites and districts, public libraries, main street merchants, antique shops, hotel/motel/bed and breakfast industry, restaurants, unique local "watering holes" (or pubs), as well as chambers of commerce.

A positive development in the area of cultural tourism worth drawing out is in the area of Native American cultural tourism. Generally speaking, the tribes in Montana practice democracy in its purest form: everyone is equal, no individual speaks for the tribe, and anyone can do business with any member of the tribe. Tribal government is an entity unto itself. However, several tribes have joined together in the Montana Tribal Tourism Alliance, a cooperative effort to promote cultural tourism in tribal areas. The Alliance is dedicated to supporting economic development of Tribal Homelands through tourism and tourism-related businesses. Through a cooperative approach, Indigenous people make decisions about which parts of their culture they will share with tourists and to what extent.

2005 marked the 200th anniversary of the Lewis and Clark Expedition, which explored the Louisiana Purchase territory and sought a water route to the Pacific Ocean. Tribal members participated in formal celebrations along the exploration route, presented the point of view of the native people who greeted Lewis and Clark, and gave some perspective on how the expedition forever changed their world.

The outlook for development through the arts in Montana

In many ways, economic development through the arts in Montana is a complicated puzzle. For example, one major piece, the State's tourism agency, Travel Montana, controls the only substantial budget for resource development—which is significantly limited by legislative mandate—yet could be more open to new ideas and working cooperatively with other entities.

The state agency, Montana Arts Council, has worked closely with another important "puzzle piece", the Department of Commerce, on how to include the creative industry in its plans. One of the public-private partnerships developed out of this was with a newspaper chain, Lee

Enterprises, which published *Montana's Cultural Treasures*, an annual guidebook and website of arts businesses aimed at the tourism market.

Another piece to our economic development puzzle is the Montana Associated Technologies Roundtable, which is supportive through information access and transfer, but is not specifically arts or creativity-related in its mission. Yet the Roundtable serves as an important information link—technology is important to the State and creativity is key to developing new technology. This link helps keeps the notion of advancing creativity in Montana as an economic resource visible with decision makers throughout the state.

In conclusion, it will be helpful when our state's leaders let go of the idea of continuing past dependence on mining, timber, and agriculture as the only economic mainstays. Instead, they should consider strategies that not only provide diversity but also come to terms with new economic realities of the twenty-first century: The Information Age.

These strategies might include how best to utilise renewable resources as well as the resident creators, makers and producers; one such approach would be greater investment in the existing wood products creative cluster. Another would be further development and promotion of tourism cultural corridors.

Moreover, creative people apparently live in rural areas because they choose to—and not because of an economic necessity that dictates this location (although cheaper housing prices help!)—and their commitment to our State should be recognised and encouraged. Governments in other rural areas comprehend the importance of creative industries and development. For example, Wisconsin has made significant commitments of staff, time, and money, and Maine has embraced the importance of resident creative industry and cultural tourism by investing state dollars in their promotion. In Ireland, resident artists pay no income tax but contribute to that country's government coffers through the value added tax (VAT).

The 2007 Americans for the Arts study, *Arts & Economic Prosperity III*—one that passed muster with noted economists—conservatively estimated that public arts activities return to local, state, and federal governments, through tax revenues which otherwise would not have been generated, at least seven times the amount all three levels of government invest in the arts through grants. According to this study, the arts and related cultural attractions are not a drain on government, but are sources of economic vitality and tax revenue that, by any measure, are an excellent return on investment. Therefore, a relatively small amount of government and private investment in creative industry and cultural tourism promotion in Montana is likely to reap benefits far beyond the initial investment.

References

Americans for the Arts. 2007. *Arts & Economic Prosperity III: The Economic Impact of Nonprofit Arts and Culture Organizations and Their Audiences*. Americans for the Arts National Report.

Diamond, Jared. 2003. *Collapse*. New York: Viking Press.

Newman, Ron. 2002. *Idea Tree*. http://ideatree.us/

Nickerson, Norma. 2005. *State of the Travel Industry in Montana*. Institute for Tourism and Recreation Research, University of Montana.

Nickerson, Norma and Thale Dillion. 2002. *Nonresident Summer Visitor Profile: A Study of Summer Visitors to Montana*. Institute for Tourism and Recreation Research, University of Montana.

Nickerson, Norma and Chrissy Oschell. 2006. 2001/2005 *Nonresident Comparsion: Visitor Profile*. Institute for Tourism and Recreation Research, University of Montana.

North American Rural Futures Institute. 2003. *The Creative Enterprise Cluster: A Montana Business Case Study*. University of Montana.

Parrish, Jose and Norma Nickerson. 1998. *Nonresident Summer Travellers to Montana: Tourism Region Report*. Institute for Tourism and Recreation Research, University of Montana.

Rademaker, Lee and Kara Grau. 2008. *Expenditure Profiles and Marketing Responsiveness of Nonresident Visitor Groups to Montana*. Institute for Tourism and Recreation Research, University of Montana.

Rosenfeld, Stuart. 2003. *The Montana Business Clusters Study*. Report for Montana Governor's Office of Economic Opportunity.

The New England Council. 2003. *Creative Economy Report*. Creative Economy Council.

Travel Industry Association of America and Smithsonian Magazine. 2003. *The Historic/Cultural Traveler*, 2003 Edition. Washington, DC: Travel Industry Association of America.

Notes

[1] The Internet's rapid development created an economic opportunity for individual Montana artists to market their work to the world. However, it is a two-way street and copyright infringement has proved an unwelcome risk; outright theft of intellectual property by pirates in foreign countries is an increasing problem for some artists. The technology that makes instant international communication possible also makes theft equally fast. For example, if an artist posts a high-quality image on a Web site, a "thief" in another country can download it, print a colour edition, and sell out before a legitimate printer in the United States can make the artist's first reproduction. Additionally, software programs exist to copy a two-dimensional image and reproduce it as a three-dimensional object. For example, this software enables a thief to copy a painting of a horse and with a three-dimensional printer reproduce the horse as model for a series of small bronze sculptures without giving the original artist a royalty or even recognition.

In effect, the artist's market for an image or object has been pre-empted and heavily damaged, if not destroyed. Although there is expensive legal recourse for an artist—if the pirate can ever be located for prosecution--in reality extremely little can be done to collect monetary damages from an international art thief. To combat these problems, artists have begun to avoid posting anything innovative on a Website, or post only thumbnail images, or post images with small pixel counts with embedded copyright notice watermarks, in order to make reproduction more difficult.

[2] www.handmadeinamerica.org

[3] www.handsofharvest.org

CHAPTER NINE

CREATIVE REGIONS IN THE DIGITAL ERA: SCREEN AND CREATIVE INDUSTRIES IN THE NORTHERN RIVERS REGION OF NSW

CATHY HENKEL

> Without a doubt, I'd call myself a screen industry creator, not a filmmaker. The next project I'm doing doesn't even involve a camera or a lens. It's being shot with a scanning laser. The movie is about the astronomy of the Maya in Central America, but the actual shooting of the movie will occur with a laser. We scan all the images at a very, very high resolution, pump them into a computer in my studio in Broken Head, edit them here, then output them to my colleagues over in a little village near Boston (they also live away from their centre in a place no bigger than Byron Bay). They have the hardware and the software to reprocess the images and send them back to me integrated for presentation…Broadband is our mode of transport and the term "film industry" is hardly relevant any more to the majority of work going on in this region.
> John Weiley, interviewed in Byron Bay, 2005

John Weiley is one of Australia's most successful and entrepreneurial producers. He moved to Byron Bay in the Northern Rivers region of NSW in 1996 and operates a lucrative international business from his home studio at Broken Head. His documentary *Antarctica* has long been one of the highest grossing films in Australia, and his work in large screen formats and utilising new technologies has been recognised and awarded in many countries. He generates most of his production finance in America and his work is exported and screened in cinemas and museums around the world. He was one of the founders of the Screen Producers Association of Australia (SPAA) and is highly regarded within the industry. His work has enabled him to travel extensively and live at various times in the UK, Europe and the USA, but he has decided that Byron Bay is "the best place on earth" to live, raise his children and operate his screen industry business.

More and more screen industry producers like John are taking advantage of the cultural and economic freedom they have earned to choose where they want to live and work. Digital technology has mobilised and dispersed production locations and major changes in telecommunications have liberated practitioners from having to reside in the cities. Increase in access to broadband and the capacity for broadband to transport vast quantities of information around the globe in a matter of minutes or even seconds means that increasingly producers like John are no longer bound to the proximity of cities to do their work.

Over the past decade there have been significant changes at every level in the development, production and post-production chain of audiovisual industries worldwide. Technological advances have been the main drivers of these changes, but so have convergence and globalisation which have had major impacts on the markets and delivery modes of content.

One indication of the impact of these changes is the increasingly widespread adoption of the term "screen industries" rather than "film" or "audiovisual industries." Screen industries are a sub-sector of the larger creative industries and are defined as individuals or enterprises involved in creating content or providing services in the following sub-sectors: film, television, print media, all digital media, photography, video, radio, performance and visual arts delivered on CD, DVD, the internet or other digital platforms. The economic value of the screen industries originates in the "intellectual property" (IP) that underpins the end product or service: the idea, the script, the photograph, design or composition. Through a process of value-added activities, the IP is enhanced and developed and may be delivered across a range of platforms or markets.

The content developed through these mechanisms is delivered in one form or another onto a screen. These include cinema or television screens, video monitors, computer screens, mobile phones and a plethora of new hand-held screens entering the marketplace. Although yet to develop in Australia, digital radio increasingly qualifies as a screen industry sub-sector as it conceivably allows listeners to select channels, read programs or get sports updates on a screen. In producing content for the screen, producers utilise the skills and services of related creative industry practitioners such as writers, musicians, performers, photographers, visual artists and designers.

Creative industries in a global context

According to Richard Florida (2005), the key factors of the global economy are no longer goods, services or flows of capital but the competition for creative talent. Florida's argument is that economic strength will in future revolve around a nation or region's ability to mobilise, attract and retain human creative talent. Economic leadership, from manufacturing excellence to scientific and technological advancement, depends on this ability.

In his 2002 book *The Rise of the Creative Class*, Florida argued that those cities and regions with diversity, tolerance, and high concentrations of creative professionals, high-tech industries and the arts are more likely to grow and be economically successful than others. Instead of people following job opportunities, companies follow the location of talent. They move to places where highly creative people have chosen to live.

In his 2005 *The Flight of the Creative Class*, Florida proposes that the new global competition for creative talent presents a significant threat to the longstanding American hegemony on several overlapping fronts. The USA, he argues, is failing to cultivate and harness the full creative capabilities of its own people in order to compete effectively. At the same time, the global competition for creative talent is hotting up. America may be losing its crucial economic advantage as the world's greatest talent magnet in favour of other global technology hotspots, which include the Scandinavian countries, Canada, New Zealand and Australia. No longer is the USA the prime source for computer software programs, games, new media and telecommunications breakthroughs, media and entertainment content or even blockbuster movies.

For example, one of the largest projects in recent cinematic history was internationally funded and crafted by the best filmmakers from around the world, but not in Hollywood. Instead the production came out of a studio in Wellington, a city of roughly 400,000 people. Furthermore, the lion's share of long-term economic benefits from the massive *Lord of the Rings* trilogy is likely to ultimately accrue not to the USA but to New Zealand.

Although Florida's work focuses primarily on urban centres and cities, the question arises as to whether the basic tenets of this thinking can be applied to non-metropolitan regions that are becoming magnets to significant numbers of the creative class. This investigation has been undertaken in a unique region in the far north of NSW in Australia. My study, referred to here as the Northern Rivers 2005 Study, has been concerned with the visibility of screen and creative industries in the region, but also with the nature of work and the needs of creative practitioners working in these sectors (Henkel 2006).

As the technology changes, the ways in which screen and creative industries are structured and operate are also rapidly changing. A "new breed of independents" has emerged that accounts for a growing share of employment and output in creative industries (Leadbeater and Oakley 1999).

These "Independents" represent a vision of the future of work with new technology that is especially appealing to the young and could influence the development of other service industries in which self-employment and micro-businesses are growing. (Leadbeater and Oakley 1999)

In 2001's *Surfing the Long Wave*, Leadbeater and Oakley argue that the new entrepreneurs in the creative industry sector are increasingly more networked and collaborative and less the "lone mavericks" of the early days of the sector. The entry barriers to most creative industries are low, and entrepreneurs exploit these barriers and contest once cosy markets. They think and act independently but will often form teams to collaborate on projects. They take risks, push barriers, try new things and are prepared to fail. They chart new pathways and are ever in search of new ideas.

Another key attribute of practitioners and entrepreneurs in creative industries is that they increasingly self-manage their careers rather than relying on employers or corporations to do it for them. They are CV and portfolio-driven and seldom stay with one company for long periods of time. A large percentage of the sector are self-employed individuals who operate as sole traders, often even without a registered business name, or in micro businesses of between one and five people. They blur the lines between employment and self-employment, full and part-time work and their work practices may well shape and determine accepted practices in the new creative economy. Most operate as contractors on a fee-for-service basis but a growing number are developing, marketing and deriving income from the intellectual property associated with their own creative work.

Leadbeater and Oakley's research in the UK is largely city based, but, as with Florida's propositions, the Northern Rivers 2005 Study considered the extent to which their research on the nature of work and entrepreneurship could be applied to creative workers in regional areas.

A brief history of creative migration to the Northern Rivers

The Northern Rivers region of NSW encompasses the area from the Clarence Valley and Grafton in the south to the Queensland border and stretches west to the Great Divide. It consists of seven local council areas and three major river systems: the Clarence, Richmond and Tweed. The region has twenty-one national parks, some world heritage listed, and is the most biologically diverse in NSW. The region is well known internationally for its two major icons: Mount Warning and Cape Byron lighthouse at Australia's most easterly point. Tourism is a significant component of the region's economy and Byron Bay in particular is a major international tourism destination.

Figure 9-1: Map locating the Northern Rivers region of Australia.

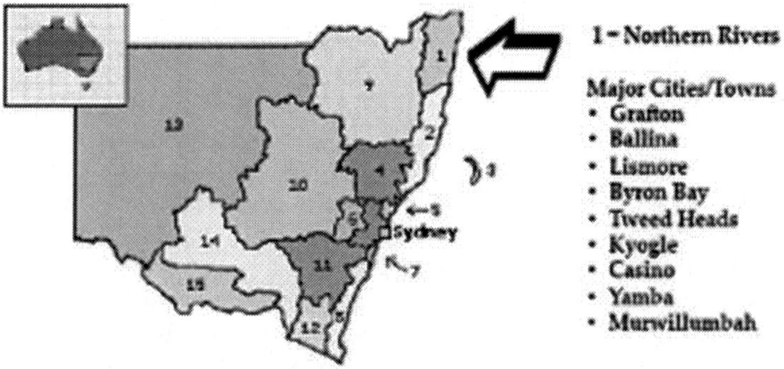

The cultural history of the region is a rich tapestry of movements. The region has been home to the Bundjalung and Gumbaynggirr people for centuries and the indigenous population maintained their cultural life through storytelling, dance and painting. Part of the mythology of early white settlement of the region is the story of a woman who came up the river on a raft with a piano. Mrs Wilson brought music and her own Anglo cultural heritage with her when she arrived with her family in Lismore in the 1850s. The Lismore Music Festival, otherwise known as the Eisteddfod, has been going for 90 years. Another regional event with a long history is the annual Maclean Highland Gathering, which has been active in some form for over 100 years.

In 1973 a major counter-cultural lifestyle event called the Aquarius Festival was held in Nimbin and it heralded a new wave of settlers looking for alternatives to the mainstream and city life. The student radicals and intellectuals who stayed on after the festival injected a rich tradition of community cultural development practice from the UK, Latin folk styles and Eastern influences in music and experimentation in visual arts and bush crafts. They set up communal dwellings and a range of "alternative lifestyle" practices, and their settlement in the hills around Nimbin, The Channon and surrounding areas undoubtedly expanded and diversified the creative arts in the region. A study conducted into this sector in 1991 identified 1,285 professional artists working in the region which was described as "the liveliest and most exciting arts area in Australia with more artists and craft workers per head of population than anywhere else in Australia" (Wynn Moylan, 1991).

Another major impact on the region's cultural life began with the opening of the Lismore Teachers College in 1970, which eventually became Southern Cross University in 1994. By 2005, more than 21% of an 11,000-strong student body were engaged in creative industry practice.

Other training organisations and providers also emerged and flourished during this period, including North Coast TAFE, the Northern Rivers Conservatorium, the Clarence Valley Conservatorium, Adult and Community Education (ACE) North Coast and in 2004, the world's largest audiovisual training school, SAE, opened its world headquarters in Byron Bay. By March 2006, the total student body in the region engaged in creative industries study or practice was estimated to be 6,220 students.

During the 1990s other migration shifts continued. Some of the leading visual artists left the region, driven away by lack of infrastructure and support, but the number of musicians, writers and performers steadily increased. There were several other influences at play affecting population shifts: improvements in transport, the increased popularity of coastal destinations, people becoming more mobile, the baby boomers beginning to retire. All these factors brought migration shifts and new arrivals from the cities and the Northern Rivers began to define itself as a "lifestyle region." Some destinations like Byron Bay, Ballina, Yamba and Kingscliff, became the focus of intense coastal development, bringing new migrants and new disposable income to the region.

The changes in technology and telecommunications that took place during the 1990s, particularly the use of email, the internet and broadband and mobile phones as forms of communication, brought another wave of new settlers to the region. Creative practitioners, and particularly people working in screen industries, discovered that they could run their micro-businesses from the coastal towns or rural areas of the region with relative ease and far lower overheads than in the cities. In 1999, John Weiley held a party for local people working in screen industries and 120 people turned up. It was becoming clear that a significant new sub-group was emerging in the region.

Scoping the local creative hub

My 2000 report, *Imagining the Future*, in documenting the size and scope of the film, video and multimedia industries, gave the region's development agencies, community and industry organisations, local, state and federal agencies some hard facts on the size and scope of local creative industries. For local practitioners, the study provided the first chance to view themselves in a connected way as an industry sub-sector.

The "hard facts" showed that at least 1,621 people were involved in audiovisual industries and a further 2,278 in related creative industries in the region; a total (conservative) estimate of 3,500 people employed in creative industries, or 4.1 per cent of the local work force. This was well above the current national figure of 3 per cent. The local sector had shown dramatic growth since the previous Wynn Moylan study in 1991 and this growth was predicted to continue. For some, the findings were dramatic and unexpected. For a handful of policy makers and practitioners, they confirmed what they already knew: that the Northern Rivers region was a vibrant and growing creative hub and a special case when it comes to creative industries.

The report showed that the creative practitioners were mainly clustered around Lismore, Byron Bay, Nimbin, Mullumbimby and the surrounding hills, and were largely disconnected from each other and any sense that they were part of an industry sector. There was one flagship performing arts company, an active writer's centre, a vibrant live music scene and a range of thriving weekend markets and festivals throughout the region, and a large number of disparate arts and community activities. Generally however, the arts community were regarded as a fringe group who contributed to the cultural life of the region and added "charm" and "lifestyle" attributes. Their economic impact was largely unknown and discounted and generally the sector was viewed as outside the mainstream; undisciplined and disorganised.

Following the release of my report, media and creative industries became priority areas for the first time in local industry and economic planning (Invest Northern Rivers 2003). A shift was beginning to take place that saw creativity moving from the fringes to centre stage in terms of the region's competitive advantage.

In 2001, a group of local screen producers formed a company called Northern Rivers Screenworks, which received both federal and state funding for a three-year industry development project from 2002 to 2005. Screenworks established a Locations Office for the region and set about implementing strategies and offering services recommended in the *Imagining the Future* report.

While practitioners welcomed the injection of funding and resources into the region, regional development agencies also wanted to know how the sector was changing, why so many new creative practitioners were moving into the region, what kind of business models they were using, what their needs and aspirations were, and whether they were finding success.

The Northern Rivers 2005 Study

In 2005 the Northern Rivers Regional Development Board published its Regional Industry and Economic Plan (RIEP) to determine priorities for regional economic development and managing projected population growth over the next twenty-five years. RIEP identified creative industries as one of ten priority areas for growth in the region and, in that context, the Development Board began work with Screenworks and Queensland University of Technology on my new study of screen and creative industries in the Northern Rivers region, published in 2006 as *Imagining the Future Part 2*.

In order to reflect the depth and breadth of the sector as locally understood, I adopted a model of creative industries depicted in Figure 9-2 below.

Figure 9-2: Defining Northern Rivers creative industries (Henkel 2006)

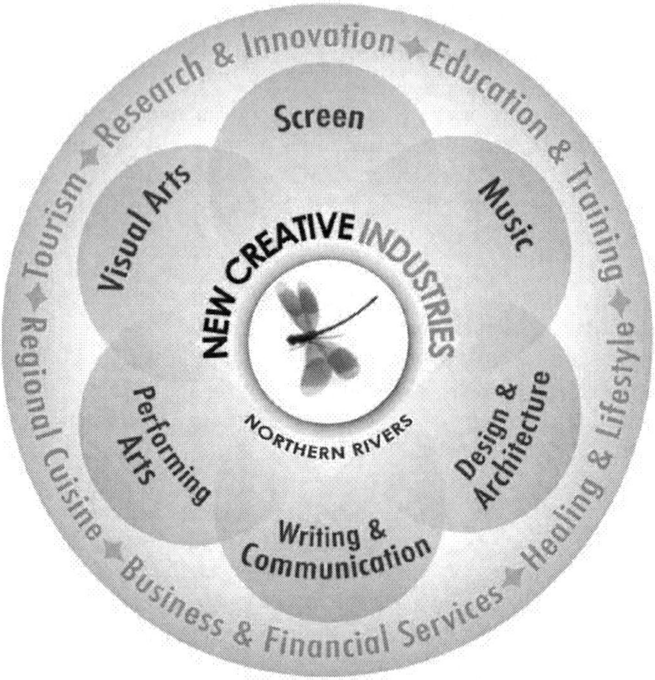

The methodology adopted involved collecting raw data from practitioners, industry groups and associations in the region, and comparing this with data collected in 2000 and ABS data for the region. This data made it possible to determine the size of screen and creative industries and to analyse the geographical spread, the size of various sub-sectors and the number of individuals and micro-businesses within the region. A survey of screen industry practitioners was also conducted across the region and a wide range of producers and policy stakeholders were interviewed. A selection of projects in development were assessed to provide a more detailed analysis of the nature of work, challenges and opportunities faced by practitioners and micro-businesses. A set of strategies for further development of the industry were drawn from the findings and these were analysed and expanded during an industry consultation tour to five major centres across the region conducted in May 2005.

The most dramatic finding of the study was (once again) the size of creative industries in the region. As of December 2005, a total of 7,280 people working in creative industries were identified—6.1 per cent of the labour force—representing a growth of more than 100 per cent since 2000. The number of screen industry producers, directors and content creators in the region had increased four-fold and those defining themselves as working in visual arts, crafts, fashion and design rose from ten to 660!

The study also found a further 1,000 people working part-time or as core volunteers in creative industries and more than 1,500 people working part-time on major events in the region. These people represented a mix of paid and unpaid work. In addition, more than 6,000 local students were studying or engaged in creative industry practice. A total of more than 16,000 local people were actively involved in creative industries.

The economic value of this sector—calculated by multiplying the number of practitioners by average earnings in this sector—was estimated at $190 million per annum, with screen industries alone worth around $40 million.

Focus on local screen industries

In addition to the data collected on the overall size of creative industries in the region, during 2004 I surveyed screen industry practitioners. The survey confirmed that the largest majority of screen practitioners work in Byron Shire (55 per cent) followed by Lismore (40 per cent), Ballina (24 per cent) and Tweed (22 per cent).

The largest percent listed their primary occupation as "writer, author or screenwriter." This was followed in order by director, producer, cinematographer, journalist and musician. More than two thirds of respondents classified themselves as self-employed, mainly as sole traders or micro-businesses of less than five people. And for some the mix of salaried, self-employed and volunteer work was complex and entwined and might change from project to project or from day to day.

The source of work was also an overlapping mix with 78 per cent of respondents listing work as being initiated or commissioned from within the region, 57 per cent from outside the region and 19 per cent nominating overseas.

The type of work being produced included making short films—rated the highest at 44 per cent—followed in order by producing commercial television, radio and print press content, education programs, documentaries, newspapers and magazines and websites.

When asked why they had moved to the region, most people nominated the area's physical beauty, lack of pollution, and climate, followed closely by a sense of community and the presence of like-minded people. The next most valued attribute was the idea that living in the region enhances personal health and provides a sense of peace, freedom and tranquillity. Another important attribute was the creative potential and developing industry of the region, and the appreciation of a culturally diverse community.

Of the disadvantages nominated, product and service gaps rated the highest. These included lack of equipment, customer service, support and skill gaps especially in the area of producing. Limited paid work opportunities rated second highest and other disadvantages included distance from cities and events, lack of skilled/professional people, lack of funding and investment and the lack of training opportunities and support.

Despite the disadvantages, 78 per cent of screen workers indicated that they were positive about the future with almost half expecting their businesses or career prospects to grow in the next five years.

Value chain analysis and flow-on effects to other sectors

The key sources of income for the screen industry sector are coming from a mix of fee-for-service project work and the marketing and exploitation of intellectual property. Greater focus on generating and marketing IP and collaboration to produce and market high value-added products and services are keys to the development of this sector.

Individual art works require marketing skills and distribution outlets.

Figure 9-3: Ecology of screen industries and flow-on effects

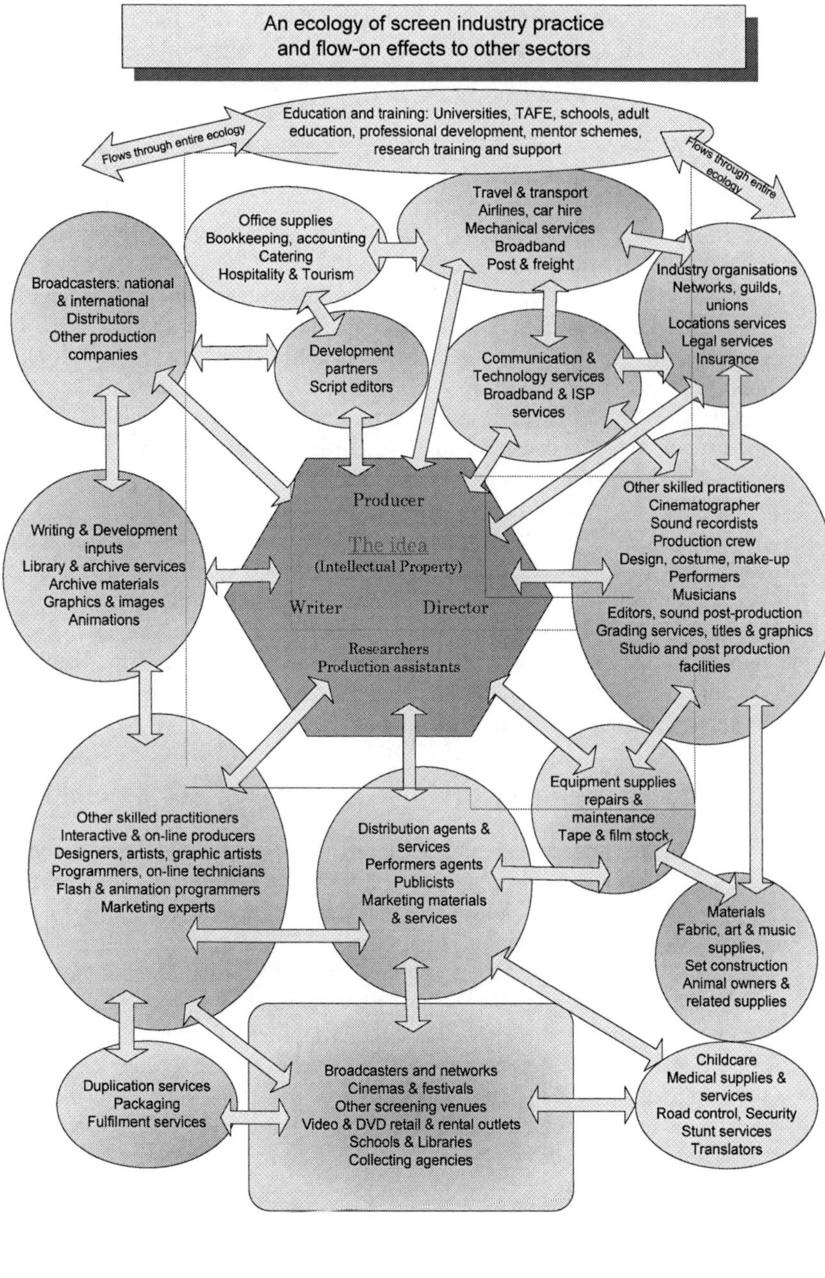

Music compositions require networked performers, music venues, radio stations and recording studios. A theatrical performance requires actors, designers, lighting technicians, promotional expertise and ticket sellers. Designs of buildings, costumes or sets require craftspeople and suppliers of materials. Journal articles, scripts or screenplays require media outlets, publishers, screen producers, cast and crew and a wide range of post-production services. The ecology of a screen industry project is depicted in graphic form in Figure 9-3 above.

In addition to the flow-on effects within the sector, there is also a significant flow-on effect to other industries and sectors. Analysis of the sector reveals a complex web of interactions with related sectors such as education, tourism and hospitality, business and equipment services, research, distribution and marketing, insurance, legal services, and even medical, safety, animal and childcare services. Understanding and developing market-focused collaboration between these sectors is vital to the growth and prosperity of not only creative industries but a wide range of related regional industries and services.

To achieve maximum market reach and deliver high quality experiences and products to audiences worldwide, the screen industries need to fully understand and utilise this ecosystem to maximum advantage. This is best achieved when the sector is viewed as a whole rather than in small parts. Skilled producers need to be networked and encouraged to form collaborative teams and to focus on niche global markets and long-tail revenue streams.

The emerging picture of a regional creative sector

The picture that emerges of screen industries in the Northern Rivers is of a large, growing and optimistic sector. There is a core group of highly experienced, credited producers generating big budget projects and employing local practitioners. Incoming productions using local locations are also injecting significant revenue into the regional economy and employing local people. The sector plays a significant role in the social and cultural life of the region and has the potential to become a major contributor to the region's economy. Screen industries have minimal requirements for land, particularly coast or industrial land, and minimal impact on the environment. Broadband is becoming the primary transport mechanism for the sector.

It is not only the screen industries that are experiencing significant changes and heightened activity. The flagship theatre company, NORPA continued to expand and diversify its program to include more music, dance, comedy and experimental works and the Northern Rivers Writers Centre expanded operations and its annual Byron Bay Writers' Festival continues to grow each year. Arts Northern Rivers—established in 2003 as a partnership between seven local councils and the state government—established a Visual Arts Network in 2005 representing over 1000 visual artists, formed an Indigenous Artist Advisory Group and partnered with the North Coast Entertainment and Industry Alliance to set up a music industry development project.

Not only have creative industries become an accepted (and respected) industry sector in the region, they have achieved "priority" status with decision-makers and have a definite "buzz" around them. There has been a significant increase in the network activity, professional development and efforts made to increase collaboration across the sectors. These efforts included proposals to develop a creative industries consortium and an arts and creative industries strategy with both community cultural development and economic development objectives.

Some intense local debate has focused around the question of whether mechanisms should support either the non-profit cultural sector or commercial initiatives, but that divide is itself misleading. The findings from my study show a picture of a "continuum" where each aspect feeds off and benefits the other (Henkel 2006). Support for the not-for-profit cultural sector creates a rich pool of raw talent, innovation and new ideas—the essential raw ingredients required by the commercial creation sectors. Both non-profit and commercial ventures increase demand for cultural and entertainment content and experiences, and both build audiences and potential markets. They are mutually reliant on each other and if they can be supported to network and collaborate, and understand the benefits of team building and relationships, the sector as a whole will benefit and grow. I represent this inter-relationship in Figure 9-4 below.

The areas of real concern for the sector include the distinct gap between experienced, credited producers and the aspirant and emerging practitioners in terms of income generation, access to markets, networks and success in generating commissions, raising finance and producing content. There are still a large number of practitioners who identify as creative industry practitioners but are earning less than $200 a week from their creative practice and are working in other areas—such as hospitality, education or small time farming—to supplement their income.

Figure 9-4: Relationship of creative industry components.

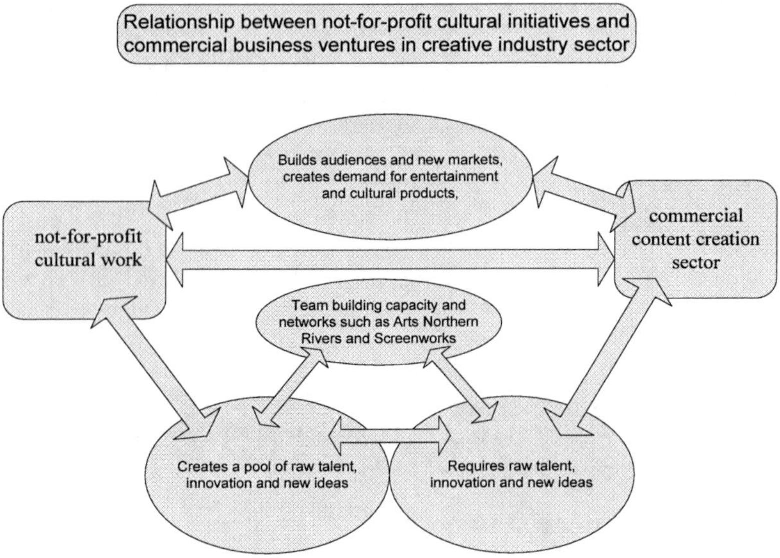

Greater access to broadband and mobile phone technology are vital for future development and there is an urgent need for more small business, marketing and promotion skills. The industry remains fragmented and needs interventions around better networking and collaboration and professional development programs. There is an urgent need for new investment opportunities and a micro-loans scheme as well as online spaces for business-to-business services, sales and marketing and the delivery of seminars and training.

There is also a need for the development of a regional collective identity for the sector and more collaboration between the industry organisations. Benefits would include a unified voice in advocacy and lobbying, the possibility of developing a collective brand, initiating major industry events and festivals and promoting the creative industries identity of the region.

Conclusions

My two studies into screen and creative industries in the Northern Rivers region not only made these sectors more visible and measurable, but also contributed valuable information on the nature of work practices and how the sector should be supported.

The 2005 study concluded that creative industries are flourishing and that there is now a critical mass of creative talent in the region. Significant high value activity has followed this influx of creative workers into the area and local producers are generating projects and products worth millions of dollars and exporting to global markets. The industry however is still fragmented and there is a need to focus more on networking and infrastructure support for practitioners, mentoring and team building mechanisms and business skills. The industry organisations in the region need further support until they find sustainable business models and the industry as a whole requires new investment and funding mechanisms particularly during the development stages.

The "new independents" operating from home offices and small studios in Byron Bay, Lismore and the surrounding hills mirror their UK counterparts as described by Leadbeater and Oakley (1999 and 2001). They come to the region seeking choice, autonomy and lifestyle satisfaction, and are willing to trade these off for insecurity and risk. Their knowledge of digital technology and global networks are their major assets and their working tools are often no more than a laptop computer, a phone line (preferably with broadband access), and plenty of hard-drive space. The major secrets to their success derive from their ability to nurture and exploit global networks, to collaborate successfully with peers and clients via email and the internet and their effectiveness at building teams and sharing skills and resources.

While there is much to be optimistic about regarding the growth of this new creative class, there are also some dangers. One of the major threats to a nation's or region's prosperity derives directly from both the expansion of global capitalism and the rise of the creative class. The tension arising from this growth was identified by Richard Florida, who argues that as regions grow more prosperous, they also grow more unequal.

> While roughly 30% of our national workforce enjoys the ability to use their creativity at work and get paid for it, they leave 70% holding on dearly to

> low paying service or manufacturing jobs, stalled on the ladder of socio-economic mobility. (Florida 2005, 17)

Within the creative centre, economic inequality can be more severe than elsewhere. For example, as the coastal town of Byron Bay has become more prosperous and increasingly a magnet to the creative class, housing has become unaffordable to middle and working class people, widening the economic divide. A new picture of Byron Bay emerges where cliques of yuppies fill the cafes, secure all the available project money and employ their mates; while buying up beachfront homes and gentrifying the villages. This "class" are blamed for increased real estate prices and the boutique shops and restaurants, leaving the "other locals" feeling displaced and angry and struggling artists from the surrounding areas feeling cynical about the prosperity of Byron and resentful about their lack of a share.

Solutions to these problems may include more affordable housing options, increasing the availability of incubator schemes and shared workspaces, enhanced networking and mentorship schemes, and more incentives for experienced professionals to offer training, nesting and employment opportunities.

Two scenarios for screen industries in the region are currently being predicted. In the first scenario, the senior producers go back to working in isolation and the question of the renewal of the industry is not addressed. Screenworks remains under-funded or ceases operations, networking opportunities reduce, cross-fertilisation between emerging and experienced practitioners declines, the leading practitioners become less engaged in the wider interests of the local industry, and the industry stagnates and fractures.

The second scenario involves collaboration between the local industry—senior and emerging practitioners—and a whole-of-government approach to developing the sector. In this scenario, all three tiers of government become meaningful partners with local industry organisations on infrastructure needs and development programs and the region continues to attract and build on its enormous talent capital. A well-resourced industry organisation—such as Screenworks—services and supports creative entrepreneurs who continue bringing resources, skills and new finance into the region. Programs are implemented to ensure the benefits of these projects and activities are spread across the region and across the various industry sub-sectors. A dynamic, networked industry emerges with a significant global export component and major economic, social and cultural benefits to the Northern Rivers region.

The needs of screen industries and the broader creative industries in the region are entwined and the necessity for collaboration and networking between the sub-sectors has been recognised by creative industry stakeholders in the region.

A formal Consortium of all creative industry organisations and training providers was established in 2006 with the objectives of advocating for the development of the sector as a whole, implementing the creative industries strategies and plans and promoting the cultural and creative capacity of the region. And in 2008 Arts Northern Rivers launched the Northern Rivers Arts and Creative Industries Strategy as a five year "blue print" for developing and marketing the local creative industries

Screen production has increased significantly in the region since *Imagining the Future Part 2* was released. *East of Everything*, a six episode television drama series commissioned by ABC television, was written and produced entirely in the region and broadcast in early 2008 and a second series was subsequently produced. My own feature documentary, *The Burning Season*, and several other screen documentaries have been produced. A high budget animated children's television series, *dirtgirlworld*—a co-production between a local and a Canadian company—went into production during 2008

It is clear that decisions taken by both industry and government at this critical stage of the region's development, and maintaining a collaborative approach, will continue to have profound impacts on this dynamic new industry sector in the Northern Rivers, and the region's future vitality.

References

Department of Employment and Workplace Relations. 2005. *Small Area Labour Markets Australia, December Quarter 2005*. Commonwealth of Australia.

Florida, Richard. 2002. *The Rise of the Creative Class*. Melbourne: Pluto Press.

—. 2005. *The Flight of the Creative Class: The New Global Competition for Talent*. New York: Harper Collins.

Henkel, Cathy. 2000. Imagining the Future: Strategies for the

Development of "Creative Industries" in the Northern Rivers Region of NSW. Lismore: Northern Rivers Regional Development Board.
—. 2006. *Imagining the Future: Part 2*. Lismore: Northern Rivers Regional Development Board.
Invest Northern Rivers. 2003. *Working Paper for the Regional Industry and Economic Plan*. New Farm: SGS Economics and Planning Pty Ltd.
Leadbeater, Charles and Kate Oakley. 1999. *The Independents: Britain's new cultural entrepreneurs*. London: Demos.
—. 2001. *Surfing the Long Wave: Knowledge Entrepreneurship in Britain*. London: Demos.
NSW Film and TV Office. 2004. *Filming in Regional New South Wales, Australia*. Sydney: Department of State & Regional Development and NSW Film and TV Office
Northern Rivers Regional Development Board. 2005. *Regional Industry and Economic Plan for the Northern Rivers*.
Wynn-Moylan, Peter. 1991. *Making Art Work: A Report into the Art Industry of the Northern Rivers Region*. Lismore: Northern Rivers Regional Arts Association.

CHAPTER TEN

CREATIVE INDUSTRIES AND CULTURAL DEVELOPMENTS IN REGIONAL EUROPE: A CASE STUDY FROM SCOTLAND'S HIGHLANDS & ISLANDS

JEREMY SIM

Context

In Europe, the distinctions between the cultural, content and creative industries are blurred. As terms used to define multiple professional disciplines of an artistic nature, "creative," "cultural," and "content" industries have large degrees of overlap and are often used interchangeably. In addition, the many nations and regions of Europe have a varied approach towards dealing with this relatively amorphous concept, although it is largely framed within the sphere of public policy. Interestingly, the strategies tend to cluster around either half of this double-barrelled moniker. Some regions focus emphatically on the "industries" part of the moniker and downplay any creative, content, and cultural merits of an activity. Others embrace creative and cultural qualifications; and blush at the thought of any commercial and industrial associations.

There are two distinct strands of discussion when it comes to the creative industries. The more established discussion is about how to best develop the core set of industries that are based on individual creativity, skill and talent, and which can reach their potential for wealth and job creation through the generation and exploitation of intellectual property. In Britain, the creative industries are taken to refer to advertising, architecture, the art and antiques market, crafts, design, designer fashion, film, interactive leisure software, music, performing arts, publishing, software and television and radio.

The other strand is less concerned about the development of those industries than it is about the impact that they have on the wider economic community and social environment. The "buzz" created by an artistic or bohemian cluster can greatly enhance the quality of life and overall desirability of a community. This idea was first popularised by Richard Florida's *The Rise of the Creative Class* (Florida, 2002), which also argued that the economically critical segment of creative professionals no longer defines itself by national boundaries but instead relocates for the best social, cultural, and economic opportunities.

The Association of Integrated Media's (aimhi) role, in this context, is as a non-profit organisation working for the development of the creative community in a remote part of Europe: the Scottish Highlands & Islands. aimhi was funded by the local economic development agency to act as a platform for creative professionals to discuss and act upon the issues relevant to working in the creative industries in the north of Scotland and to build local-global partnerships.

Creativity: The Supporting Role

Florida's bestseller, released five years after the creative industries were first popularised as a policy concept, drew attention to the positive spillover effects that creative professionals have on their economic environment. This assessment of the creative industries is less about individual production and performance than it is about improving life for everyone.

Everyone likes to make headlines for positive reasons. For policy makers, there are few headlines more positive than those that trumpet job creation. Since it is easier to create great numbers of low-paid jobs than high-paid ones, the common model for job creation has historically been based on chasing low paid jobs for low skilled workers. The unintended result is the prioritisation of quantity of employment over quality. As recently as a decade ago, the favourite strategy of economic development managers was to provide huge subsidies or tax breaks to companies for building new factories, mines or agricultural compounds in a given area. In return, the companies would create a large number of low-paid jobs, which often lasted only until the subsidies or the primary materials were exhausted, or until a better deal was offered by some other economic agency elsewhere.

However, in the global, networked economy, high-calibre professional talents also possess the level of mobility and selectiveness normally associated only with fickle multinational corporations. Today, it is the high-quality jobs that follow the talent and not the other way around. Talented and capable individuals who can live and work almost anywhere in the world will naturally gravitate towards locations that offer them the highest quality of life. Thus, when it comes to the very best human capital, aesthetics do matter. An attractive living environment can draw individuals of high calibre with commensurate earning potential. Whether it is the mountain sports or the ocean view, an avant-garde cultural life or the cafés and quality of coffee, the perceived "buzz" and "liveability" of a place is an important factor in deciding where in the global value chain that region will occupy in the long run.

The real estate business in California provides a simple analogy to illustrate this phenomenon. There is a property investment strategy based around the central tenet that "where the artists go, the yuppies will follow." Canny property development companies look for rundown or working class neighbourhoods where artists, students and hippies have recently moved in. The company will buy up property in this neighbourhood, at first only a handful at a time, and then *en masse* as soon as there are funky galleries, bookshops, organic groceries and cafés in sufficient numbers. These properties are then profitably sold off to "yuppies," "dinkies" and other white-collar professionals attracted by the "buzz."

Creative Industries Network Models in Europe

Outside of the British Isles—where the term "creative industries"—was coined, there are a few countries that should be singled out for their work in harnessing the creative energies of their population to produce economic results. From Albania to Sweden, the policy makers of just about every European country have at least some interest in the creative industries. This part of the section deals briefly with two of the more striking models.

It should not surprise anyone that the Scandinavian countries are amongst the world leaders when it comes to the industrial application of creativity. Denmark and Sweden are particularly strong in the use of design in just about every facet of the national life; although Finland, Norway, and Iceland are no laggards in this department either. The seemingly ubiquitous talent to combine functionality and aesthetic in those countries is a tremendous asset, and one that lends them a clear advantage over the rest of the world. These Scandinavian countries also excel in collaboration and teamwork.

The Jenka Creative Industries Network draws together leading Scandinavian institutions and organisations from the creative industries, specifically within the fields of art, culture, education and research. Jenka sets itself up as primarily a "dating agency," with the core aim of connecting organisations in the belief that innovation and development evolves from unconventional experiences and strong personal relations. Since the creative experience tends towards the original and collaborations are significantly influenced by personal relationships, the Nordics have opted to put in place processes to matchmake creatives and leave it to individuals to develop their own partnerships rather than for more prescriptive measures.

The Network differs from aimhi, in that it caters mainly to large research and public sector organisations and is more focused on public policy advice than practical industry development. However, Jenka does offer a general matchmaking service where organisations can submit calls for project partners. The Network operates in a similar way to aimhi in that both seek to connect creative professionals—so as to catalyse creative collaborations. The main difference is that Jenka has a larger catchment area; connecting all the Nordic countries. This enables them to include a greater variety of creative organisations: from research institutes to multinationals to freelancers.

Another interesting model for creative industries development can be found in Austria where their Creativ Wirtschaft initiative, by being located within the Federal Economic Chamber, has a greater global reach powers than "chambers of commerce" elsewhere in the world. The Chamber coordinates and represents the interests of the Austrian business community on the national and international level. As membership is compulsory it comprises all Austrian companies. The organisation is financially self-sufficient and self-governed, making it independent from government. The Federal Economic Chamber's reach and access to private capital and initiative makes it one of the most commercially oriented entities in the development of the creative industries. Being part of this powerful business lobby, Creativ Wirtschaft is well placed to address and influence policy makers about the importance of creative industries for civic and economic wellbeing.

The creative policies of Nordic Europe and Austria are geographically inclusive. This means that—theoretically at least—rural areas are treated similarly to the urban centres. In the case of Nordic Europe, a region with Europe's lowest density of population, most of the territory is "rural" and "remote" by definition. Of course, the challenges of working in remote areas means that it can be harder to make those professional and commercial connections that are easier to come by in metropolitan areas. One example of overcoming this major disadvantage is the Integrate to Innovate (i2i) initiative that aimhi lead. Funded by the European Union's Northern Periphery program, i2i is a public-private-academic sector collaboration spanning Scotland, Sweden, Iceland, Finland and Norway. This million-euro project began in 2005 to 2007 with the purpose of studying how human networking, facilitated by technology, can promote innovation and creativity in the most remote and rural fringes of Europe.

Attracting the "Best" and "Brightest": migration within the EU

In recent years, the integration and enlargement of the European Union is one of the single biggest macroeconomic factors affecting Britain and, in particular, Scotland's Highlands & Islands. At a time when most industrialised countries are undergoing a population decline, the dynamic flow of human capital within the EU provides a unique opportunity for European countries to reposition themselves as prime destinations for the talented and entrepreneurial to work and live. It is not only a chance to attract migrants from fellow EU countries; it is a chance to attract the best and brightest from those countries.

In many ways, Britain (and Ireland) has been winning this race hands down. According to a 2007 report by the Ernst & Young ITEM Club, immigrants from the new EU states are helping to keep inflation and interest rates down, boosting consumer confidence and house prices, and their tax contributions are helping ease Britain's pension problem. The EU's two native English speaking countries have managed to put in place legislation facilitating a free flow of productive migrant labour while apparently discouraging would-be "freeloaders." For example, while the UK and Ireland opened up their labour markets fully to the ten new accession states in 2004, migrants from those countries were not eligible for state benefits until they had been in residence for two years. The policies were not without their critics, but the fact is that the successful balancing act has helped the UK and Ireland reap the continent's fastest population and economic growth rates.

In his article in *The Times*, 30 April 2004, "How ten new members can help us change Europe," British Prime Minister Tony Blair defended the UK's open door policy towards the new states from the 2004 EU expansion. "Tomorrow's expansion to the East and South," he wrote, "is good for Europe and good for Britain. Good for our security, for our jobs and prosperity."

And most other EU countries—with the notable exception of the Netherlands and parts of Scandinavia—are missing out on this opportunity. Faced with the dual threats of rising unemployment and an exaggerated perception of cheap East European labour as "dole seekers" flooding the West, much of Western Europe lurched to the Right. By allowing xenophobic paranoia to bully them into a bunker mentality, many countries have missed out on the potential benefits. Studies in the US show that the average migrant from a source country tend to be more skilled and motivated than the average non-migrant from the same country (Borjas, 1991; Chiswick, 2000). Those findings are supported in the UK's experience. A 2005 report from the Institute for Public Policy Research, *Paying their way: The Fiscal Contribution of Immigrants in the UK*, showed that migrants paid approximately £41 billion in taxes and that the migrant workforce, which makes up 8.2% of the population, pays 10.2% of all income tax collected. Separately, UK ministers estimated that migrants contributed an estimated net gain to the economy of £2.5 billion after taking into account the added cost on public services.

As of the end of 2005, London and historically prosperous southern England were bursting at the seams. Most of England can seem like one huge continuous urban development for those used to the open spaces of Australia or North America. More remarkably, the Scottish Highlands & Islands, a region of Britain that has experienced centuries of population and economic decline, has undergone a dramatic reversal in recent years. The region now boasts close to full employment and Inverness, the largest town, is amongst Europe's fastest growing in terms of population. Most of the new migrants to the Highlands are urbanites from Britain or the Continent, often in search of wide-open spaces and a quality of life that is hard to find in most of Europe.

Of course, the Highlands & Islands have been an exclusive holiday destination for royalty, the rich and the famous for centuries; but now the region has also become home to the technologically savvy and the professionally mobile. Certainly, few who have ever visited the Highlands—or Iceland or Norway—would argue that these remote regions are not amongst Europe's most spectacular landscapes. There is a reason why the Highlands & Islands has become a popular backdrop for filming, from commercials for Toyota to the blockbuster Harry Potter franchise.

Newcomers might be drawn to the Highlands & Islands because of the scenery, but they usually end up staying because of the people who are already here. Highlanders are famously hospitable and tolerant and very proud of their heritage. In addition, there is a deep vein of cultural riches that is not immediately apparent to the casual tourist. A vast oral tradition is arguably the Highlands' greatest cultural treasure, one that has materialised in the forms of innate musical talent and verbal agility, as displayed in the locals' "gift of the gab" and an infinite store of myths and stories. If the British have the world's best sense of humour, then the Scots are the most instinctively funny people in Britain.

The "celidh" is one of the highly flavourful traditions handed down over the centuries. At its cosiest and very best, groups of friends and neighbours take turns to host a celidh at their homes. Alongside much drinking and eating, guests must each take a turn in entertaining the assembled group. You can tell a joke, sing a song, tell a local story or show off a traditional dance-step. The audience is a forgiving and enthusiastic lot, but everybody *must* take a turn. Picture the snow lying on the ground outside, a wood fire roaring in the fireplace and a couple of drams sliding hotly down your belly. The entertainment and the sheer warmth of the hospitality make for a rare treat for anyone taking part.

The inflow of migrants into the Highlands & Islands has in turn contributed to an increasingly cosmopolitan and diverse environment. The area is not yet, and will never be, Southern California. However ethnic restaurants, organic groceries, funky music stores and bookshops are beginning to mushroom across the region. So too is the number of alternative lifestyle options, from tai chi classes to a gay community. Thus far, simple common sense on the part of both the native communities and the incomers, and especially local government agencies, has ensured that the burgeoning social diversity has not come at the expense of the rich indigenous culture, or vice versa.

To be sure, the recent immigration has lead to rising house prices, but this is moderated by the fact that many locals have benefited from the economic growth. In any case, the popular Scottish view has been that rising house prices is a small price to pay for getting "new blood" into a sparsely populated region with a declining birth rate that has traditionally been a region of net emigration.

Since the EU expansion of 2004, a large number of Polish people have arrived in the region, making them now the second biggest minority group after the English. In some towns Polish communities have gone from being non-existent to being more than 10% of the population. Yet I have heard nothing that was not overwhelmingly positive from the local population.

In 2005 the Scottish Government implemented a Fresh Talent Initiative to encourage talented foreigners from Asia, North America and the rest of the world beyond the EU to come to Scotland to study, work, live and settle. In addition, high profile media campaigns by the Scottish Government such as the "One Scotland, Many Cultures" initiative have reinforced Scotland's well deserved reputation as the most pro-immigration region in Britain. In one advertisement, it was pointed out that the Scots have been emigrating for centuries and that it was only fair to welcome immigrants to Scotland the same way a Scottish migrant would want to be welcomed into Canada, Australia, Spain or Hong Kong.

However, attitudes towards migrants have been less positive in the south of Britain, where large scale immigration has been ocurring for a long time.

Businesses have been the most vocal supporters of immigration. In an official statement in 2007, The Confederation of British Industry said, "Migrants to the UK bring valuable skills and ideas with them and help to fill job vacancies where Britons are unable or unwilling to do so."

The public opinion is a different story. An Ipsos poll for *The Sunday Times* of August 20, 2006, "Public wants much harsher immigration policy, says poll," found that only 14% of people polled agreed that immigration is "generally good" for Britain. Nearly half of those polled disagreed that allowing foreigners to settle in Britain is good for the country and 63% thought immigration laws should be "much tougher."

Creativity's Leading Role in Scotland's Highlands & Islands

Without downplaying the benefits that creativity and cultural activity brings to the whole community, in this section I will focus specifically on the development of the region's creative industries from a commercial perspective.

In many ways, Britain and the Highlands are made for the creative industries. This sector is already huge in the UK, contributing four times more to the national economy than agriculture, fisheries and forestry combined and with exports growing at a 13% in 2000 (Department of Culture, Media and Sport, 2001).

London is the world's most popular visitor destination and, with popular media and entertainment globally dominated by the English language and Anglo-culture, few would question London's position as Europe's entertainment capital. However, even the charitable might query the suitability of an attempt to position the Highlands & Islands for the creative industries.

There are three reasons why the Highlands are a good fit for the creative industries. The first is the appeal of the inspirational environment to creative minds—landscape and a sparse population make the region a non-urban alternative to London. The second is that the potential obstacles in working in the Highlands—chiefly geographical remoteness and physical distance from major markets—are least likely to affect the technologically savvy and nimble small companies that dominate the creative industries. And, finally, the region has a long tradition of innovation, improvisation and originality.

The popular image of the Highlands & Islands as a picturesque backwater, with stunning scenery and little else, is no longer true.

Yes, there are less than half a million people in a territory the size of Belgium.

Yes, the pace of life is slower than London or Edinburgh.

It is also true that there is little in the way of manufacturing or services industries here—whisky and tourism dominate the local economy.

However, the Highlands & Islands have always been home to the bold and pioneering, the eccentric and quirky. As a result, there are niches of true excellence scattered about. For example, with just 7% of Scotland's population, the Scottish Highlands are home to half of Scotland's Michelin starred restaurants. It might also surprise many that the Highlands not only have some of the highest national literacy rates but that the spoken English in the Highlands is so clear and prized that Highlanders are highly sought after for radio and television work throughout Britain.

In this remote and rural part of Britain, multi-jobbing—or "cyber-crofting" as we call it in the Highlands—is common. It is not unusual to come across advertising copywriters who also do a bit of farming, television scriptwriting and play the bagpipes at weddings. The traditional Celtic quartet playing at the local arts festivals for free may also well be trying to put together a touring gig and a record deal. aimhi understands that a significant part of many creative professions is about multi-jobbing, or juggling projects that are either short-term or responsible for only part of one's income, or both. In contrast with a secure, permanent job with set routines, multi-jobbing requires a different mindset and work ethic. In fact, there are some similarities between the traditional multi-jobbing you see amongst Highlands' blue collar workers, and the creative multi-jobbing you see amongst local creative professionals.

An unusual experiment in developing creative industries has taken place in the Highlands. Within the region, the local economic agency, Highlands & Islands Enterprise (HIE), sponsored three main groups in delivering both non-profit "pure" arts services and for-profit creative content ventures. Hi-Arts has a strictly non-profit brief in dealing with arts and cultural projects. The other two agencies have a much more commercial outlook: Highlands & Islands Record Labels (HAIL) deals specifically with music, and The Association of Integrated Media-Highlands & Islands (aimhi) with general creative content.

With culture regarded as both a commercial and social good, in addition to the large grey area between explicitly non-profit cultural initiatives and explicitly commercial creative content projects, the non-profit cultural activity is looked upon as a pool of raw talent that the commercial creative sector can tap into. So there is a great deal of collaboration between Hi-Arts and two more commercially focused agencies. For example, all three bodies co-organised *Go-Hi 2006*, a showcase of the creative industries in Inverness which was attended by record label representatives, Channel4 executives, and literary agents from all over the UK.

I have observed in some countries the attitude that a commercial approach to (or even association with) creativity may risk "critically tainting" aesthetic integrity. An explanation of why the situation is different in the Highlands & Islands is that the non-commercial arts are better funded in Europe than anywhere else in the world, where it is more readily accepted by taxpayers that public investment in commercial creativity does not come at the expense of the non-commercial sector. Another explanation is that the Scots have observed how their Irish cousins have prospered by slickly packaging Irish culture and creativity to the world, and how that has benefited Irish society.

The most likely explanation however is the long local tradition of mutual dependencies within and between the public and private sectors, and the preference for communal initiatives. In such a small community, any initiatives tend to involve many of the same faces and would be challenging to make progress without the cooperation of those individuals. Communal initiatives tend to be inclusive and developed with care that they strengthen (not weaken) local communities.

In any case, awareness about the commercial potential of the unique Highlands culture has only reached a critical level in the last few years, and the experimental nature of current policies reveals that we are still trying to find the right mix.

The first experiments in commercial development came in the music sector. It made sense, as the region had a vibrant and well-established traditional music scene, which had been developed through the work of community-based organisations, such as the Fèisean, who support and encourage traditional music. With support given also to develop artists playing more mainstream contemporary music, there has traditionally been a plethora of talent, but the mechanisms for distribution were weak. HAIL was initially established to support the record labels in the region, but the organisation's remit changed in time to supporting the music industry as a whole. In 2000, when HAIL began, there was one local, commercially-inclined contemporary music act. By 2006 there were more than fifty. In the years between HAIL put on music and marketing workshops, set up industry showcases, and facilitated touring gigs in Europe and North America for local bands.

HAIL's brief can be summarised thus: create access to networks that will support the music; create new national and international partnerships that provide access to new markets and create the scale that would have been hard to achieve in the Highlands & Islands alone; and, finally, provide management support to artists and businesses. HAIL's annual showcase, *goNORTH*, now attracts musicians from the UK and Scandinavia and record label executives from around the world. The foreign acts top up the local talent pool on display; making it worthwhile for record label executives from New York and London to make the journey to Northern Scotland to scout for talent. Recently the event has also added sessions on screen industries, designer fashion, writing and publishing.

The success of HAIL can be seen in the growing recognition of the region by the wider industry, the number of new artists being signed, the development of new growth, albeit small, of management and other businesses, and finally in the increasing success of festivals and live music in the region. One festival alone, the Rock Ness Music Festival, staged (of course!) on the shores of Loch Ness and headlined by Fat Boy Slim and Razorlight, pulled in 35,000 concertgoers in 2008 and pumped millions into the local economy. The Festival was organised by private concert promoters whose businesses were themselves originally developed with financial and business assistance from HAIL.

aimhi is an eclectic grouping of about 130 creative companies; ranging from television to video game design and from publishing to design. The size of the companies ranges from a single freelancer to 40 employees or more. aimhi networks local companies and also teams them up with potential collaborators of high calibre from outside the region.

In early 2005, a Highlands & Islands Enterprise-commissioned audit of future growth industries in the region threw up surprising results for the creative industries (See Tables 10.1 and 10.2).

Table 10.1: Indicators for Future Growth Among Key Industries in Scotlands Highlands & Islands (Highlands & Islands Enterprise 2005).

Future Growth Industries	Average Salary (£)	Projected Employment in 2013	Number of Companies	Investment Readiness	Public Sector Investment (£ millions)
Creative Industries	27,000	1000	750	Medium	0.62
Renewable Energy	20,000	1500	125	Low	21.1
Nuclear Energy	24,000	1000	25	Medium	0.77
Healthcare & Biotech	36,000	1500	13	Medium	15
Marine Science	42,000	500	7	Low	4.75
Information Technology	25,000	1000	620	High	1.4

Table 10.2: Demographics of creative professionals in the Highlands & Islands (Highlands & Islands Enterprise 2005).

Criteria	Other Growth Industries	Creative Industries
Percentage with degree or more	33%	40%
Percentage who are new migrants	27%	40%
Percentage who are export focused	51%	85%
Percentage who are female	48%	71%
Percentage who are lifestyle businesses	27%	25%

The creative industries outperformed the rest of the growth industries in the key areas of postgraduate education levels, attracting new migrants, export focus and equal opportunity. This is in spite of the disproportionately tiny amount of public investment the creative sector has received in comparison to these other two sectors

We haven't as yet expended much analytical effort trying to understand why local creatives have enjoyed such success. My gut feeling is that places like the Highlands & Islands, which is rich in cultural life and natural beauty, have always attracted the artistically and creatively inclined. However, we have not always been able to take advantage of this "pull factor"—and that is something aimhi wants to change.

So, in spite of a small talent pool and historically a low priority amongst policy makers, the Region's creatives have done remarkably well. Since 2005, Highlands & Islands companies have notched up kudos in creative sectors as diverse as television, music, design, publishing and corporate communications. Platform PR, a small local firm, beat out corporate giant Weber Shandwick for the 2005 Best Public Relations Company Award in Scotland. And in the same year, Canan, a publishing and multimedia company, won the Department of Trade and Industries' Best Technology Transfer Award over some of the world's leading multinational biotech and engineering companies. A local artist with no prior television experience went through an aimhi television production workshop and made his debut film for Channel4 six months later, then garnered a BAFTA nomination for Best New Talent.

But my favourite success story must be about the collaborative effort between a dozen local Highlands companies and freelancers who pulled off a superb advertising campaign on a scale normally associated with much larger ad agencies. Not only did they collect national advertising awards for their project, they also showed us that David could beat Goliath, especially if the Davids of this world can join forces.

The challenge has been, however, the lack of depth in this talent pool, a direct function of the small creative population. Two possible solutions were identified. One approach has been to bring the positive qualities of the core creative industries into the wider Highlands & Islands business community, especially amongst industries that would disproportionately benefit from an increased amount of creative input. For example, a series of workshops were developed for the tourism industry specifically to enhance the interactivity of tourist attractions using processes commonly found in film, photography and literature. The core of a filmmaker's—or any creative content producer's—work is to find a compelling story and communicate it effectively to an audience. The tourism business, although it is not always aware of it, is not very different. In 2006 at a creative tourism event, aimhi's *Style Talent Development* seminars worked with hoteliers and tour operators to develop new tourism products.

The other tack that has been adopted by associated organisations like

HAIL is to build alliances with likeminded groups from other peripheral areas of the world. New technologies have allowed interactions, collaborations and a trade in opportunities with talented creatives globally.

Conclusion

The best thing about for creative industries and the electronic age in which we live is that traditional barriers such as distance from consumer markets are less of a problem for many creative practitioners. Designers and filmmakers can more easily live and work in the Highlands & Islands than they might have even a decade ago. Advances in modern communications technology means that the "virtual hub" is slowly eroding the absolute need to be located near a geographical hub. There is still a great advantage associated with being located in London—proximity and ease of physical meetings with agents and representatives for example—but at least it is now conceivable to have a thriving creative practice while living full-time outside of the big metropolises.

The spectacular growth of online content platforms like YouTube and MySpace are matched by a rapid decline in profitability amongst the giants of traditional publishing, broadcasting and record label sales. This phenomenon spells "opportunity," not only for the teenage entrepreneur in her or his LA or London bedroom, but also for creative professionals living in rural and remote areas such as the Highlands & Islands.

The European Union's Northern Periphery Program (NPP) has allowed Scotland's Highlands & Islands to work closely with regions that share similar geo-physical conditions. Being a NPP region means sharing experiences with similar remote regions in Iceland, Greenland, Norway, Sweden, the Faroe Islands and Finland. The i2i initiative is an example of how the EU can help to network and link these regions in order to boost local innovation and creativity.

And evidence is abundant that good networking promotes increased business-to-business activity, as well as opportunities for creative thinking and problem solving. This is particularly important in Europe's Northern Periphery region given the climatic, geographic, environmental and social composition.

Perhaps the biggest challenge for those involved in the development of the creative industries in the Highlands & Islands is in successfully balancing the patience needed to develop an industry with the energy necessary to keep the development initiatives rolling. In this, we have drawn upon the experience of another vital local industry. In the manufacture of single malt whisky, one of the most important steps is to collect the distillate at the right time. Collect prematurely and the low-boiling compounds make the spirit unfit to drink. The desirable "middle cut" contains fruit and flower-smelling esters, along with optimal concentrations of complex oils. Then you allow the distillate to age in a wooden cask and it is this maturation process that gives a prized single malt its depth, complexity, and flavour. The creative business is not so different.

References

Borjas, George. 1991. Immigration and Self-Selection. In John Abowd and Richard Freeman eds. *Immigration, Trade and the Labour Market.* Chicago: University of Chicago Press.

Chiswick, Barry. 2000. Are Immigrants Favourably Self-Selected: An Economic Analysis. In Caroline Brettel and James F. Hollifield, eds., *Migration Theory: Talking Across Disciplines.* New York: Routledge.

Ernst & Young ITEM Club. 2007. *Migration and the UK economy*. ITEM Club Special Report, September.

Florida, Richard. 2002. *The Rise of the Creative Class. And How It's Transforming Work, Leisure and Everyday Life.* New York: Basic Books

Highlands & Islands Enterprise. 2005. *Creative Industries Audit: Indicators for Future Growth Among Key Industries. Inverness.*

Sriskandarajah, Dhananjayan, Laurence Cooley and Howard Reed. 2005. *Paying their way: The Fiscal Contribution of Immigrants in the UK.* Institute for Public Policy Research.

Department of Culture, Media and Sport. 2001. *Creative Industries Mapping Document-2001.*

PART IV:

COMMUNITIES AND WELLBEING

CHAPTER ELEVEN

COMMUNITY LIFE AND CULTURAL VITALITY IN A CHANGING WORLD: VALUING GOOD PRACTICE IN COMMUNITY ARTS

MARTIN MULLIGAN

Introduction

It is a vital sector of the arts "industry" in Australia but there is little agreement on what to call it. It used to be called "community arts," but much of what was done under that name was seen by most arts practitioners as being low quality art or, at best, art for passive leisure. Around thirty years ago the Australia Council established a Community Cultural Development Board (recently abolished) that was initially aimed at giving more Australians the opportunity to access and/or participate in arts activities (Mills and Brown 2004, 6). Over time the emphasis changed to one of helping communities develop projects and programs that could express their own cultural identities (ibid). However, many practitioners think the term "community cultural development" is a bit prescriptive and ponderous and they probably feel more comfortable with the old "community arts."

A more recent drive to link the arts to economic productivity has been aided by the work of Richard Florida who wants us to talk about the "creative industries" and a "creative class" of people who can use their creativity to give leadership to communities ranging from the local to the national (Florida 2002 & 2005). However, Florida's elitist notion of creativity resonates more with people in government and business than it does with community arts practitioners.

A former director of the Community Cultural Development Board at the Australia Council, Jon Hawkes has argued that it is important to distinguish between "arts policy" and "cultural policy" with the latter embracing the wide range of ways in which communities engage with arts practices (Hawkes 2005). He has pushed, quite successfully, for the idea that cultural activities at a community level should be more thoroughly integrated into processes of planning and governance (Hawkes 2001). In his attempt to mainstream community arts he is less interested in what practitioners might call themselves than in the impact of their work on the "cultural vitality" of communities, and this is probably one way to vitalise the discourse. However, research that the Globalism Research Centre at RMIT University has conducted for the Victorian health promotion agency VicHealth on links between arts, celebrations and community wellbeing suggests that there are some extraordinary practitioners of community arts in Victoria who feel that they get little external acknowledgement for their work. They feel that there is little encouragement for people who have acquired advanced skills in this work to continue working in the field. (Mulligan et al 2006)

People such as the director of the acclaimed Melbourne-based Rawcus Theatre Company, Kate Sulan, and visual artist Giz James[1]—who for a number of years ran art classes for people with psychiatric "disabilities" in the St Kilda area—say that their artistic engagement with marginal sectors of the community is also good for their own art but they feel that there is still a rather snobbish dismissal of the work of community-engaged artists. Of course, there are very few "professional artists" in Australia who can make a comfortable living from their craft (see Throsby and Hollister 2003). Arts economist David Throsby has been very prominent in making the argument that the arts deserve greater government support yet he has only recently broadened his concern from arts policy to cultural policy (2006) and in doing so he has become a stronger advocate for the neglected community arts.[2]

Hawkes and Throsby are probably in agreement that "cultural policy" should not be aimed at imposing particular conceptions of Australian identity, or even notions of what might constitute "the good life," on local communities. Rather it should aim to give expression to diverse representations of contemporary experience so that we might have a more vigorous and inclusive debate about where we are heading as a nation. Arts practices at a community level can open up the discourse about the values and aspirations that can help Australian society become more tolerant and innovative at both local and national levels. At the same time the term "cultural vitality" probably helps us see the importance of the link between arts practices and community wellbeing. Given that VicHealth has undoubtedly been at the forefront of Australian efforts to invest in community wellbeing as a primary health strategy, the research conducted for VicHealth by the Globalism Research Centre will be of interest beyond Victoria.

The Globalism Research Centre study included an examination of innovative community arts practices across four very diverse Victorian communities; those centred on inner-urban St Kilda, outer-urban Broadmeadows, the popular tourist town of Daylesford in the Central Highlands, and the region centred on Hamilton in the state's western district. Research methods included an innovative use of survey techniques, lengthy case-study interviews (called "strategic conversations"), a collection of community stories, and the use of photo-narrative techniques to get a broader range of perspectives on what constitutes community life.[3] The study certainly provides evidence to support investment in community arts (broadly defined) to improve or maintain health and wellbeing. However, it also suggests that community arts projects and programs need to be well targeted, well designed, adequately resourced and carried out by skilled practitioners in order to have good wellbeing outcomes.

Global-local challenges

As the name suggests, the Globalism Research Centre works on the premise that local communities cannot avoid an active engagement with global processes and influences. However, most studies on the impacts of globalisation are focused on developments at an international level and their impacts on nation-states and supra-regions. Contemporary globalisation is most often equated with capitalist economic expansion and associated neoliberal ideologies (see Steger 2005). This has been dubbed "globalisation from above" (Falk 1999; Nairn and James 2005). Local communities are usually seen as having to adjust and fit in with changes that are occurring at "higher' levels. Two recent international studies (Savage et al 2005; and Brennan 2003) have made a start on reversing the emphasis, but the field is still thin and under-researched. Given the dispersed population and the centrality of the twin concerns of regionalism and communications to successive Federal and State governments, surprisingly little work in this regard has been done in Australia.

Furthermore, in the context of the new "fluidity" of global interchange, the very notion of "community" has come in for sustained critique in relation to its effects on social wellbeing. The feminist scholar Iris Marion Young (1990 and 2000) has long argued that the desire for community is oppressive in that it seeks to nullify difference and, as a consequence, validates gender discrimination, racism and ethnic chauvinism. Similarly, Zygmunt Bauman has argued that communitarianism creates an ideal of community that is like the "home writ large" in which there is no room for the homeless and which can also turn into an unexpected "prison" for many of the residents (2000,171-2). Bauman is more optimistic than Young in thinking that a new kind of unity is possible—"a unity put together through negotiation and reconciliation, not the denial, stifling or smothering out of difference" (2000, 178). However, under conditions of what he calls globalising "liquid modernity", he sees community as entirely a matter of individual choice—a desire to redress the growing imbalance between individual freedom and security. Arjun Appadurai has gone further in saying that he far prefers the "detachment of a postcolonial, diasporic, academic identity" to the "ugly realities of being racialized, minoritized, and tribalized in my everyday encounters" in local community (1996, 170), and he looks forward to the time when externally and internally oppressive nation states will give way to "transnational social forms" (1996, 177). In all these cases, community-as-such is seen to be source of the problems of exclusion and conflict. It is our contention, however, that the theorists of this "postmodern fluid world" fail to understand the enduring, if changing, possibilities of existing communities—from the local to the national.

Rather than projecting visions of a cosmopolitan imaginary, it is arguably just as important to revisit the principles that frame how we live together in real and imagined communities ranging from local and regional communities to the national and the global. In a recent book, Globalism Research Centre founder Paul James has argued that globalisation requires us to think much more deeply about how "social relations should be based ethically and practically on positive principles of interrelationship"; principles that "need to be argued about, negotiated and worked through the various modes of practice" (Nairn and James 2005, 118). Rather than fixing on the oppressive nature of social exclusion as many studies do, James argues that we should begin from the premise that we need to work through the tensions associated with practices of exclusion and inclusion, freedom and obligation, difference and identity, and authority and participation. There are no obvious answers here, but approaching local community in this dialectical way is more likely to elicit viable pathways to sustainable regional development than the dominant "developmentalist" programs.

Research by the Globalism Research Centre on the links between arts, celebrations and community wellbeing (mentioned earlier) has suggested that a range of excellent community arts projects in local Victorian communities is helping those communities to rethink the nature of contemporary community life (Mulligan et al 2006). However, it is important to understand that this work is not specifically local in nature because it reflects an Australian community arts tradition that has had many global influences.

The emergence of community-engaged arts practices

That community arts began to take on a community "empowerment" agenda in Australia during the 1970s is no surprise given the growing influence at that time of various "social movements" advocating "social justice" for women, Aboriginal people, young people, and "ethnic" minorities.[4] The broadly political movement for the emancipation of "oppressed" sectors of society drew inspiration from the writings of people involved in similar struggles in other parts of the world. A highly influential figure was the Brazilian adult educator Paulo Freire whose 1968 book *Pedagogy of the Oppressed* became a must-read for social "activists" in Australia after it had been translated and circulated in English-speaking countries in the early 1970s. Freire's pedagogical framework for working with the oppressed—called "conscientisation"—was taken up by the Brazilian community theatre worker Augusto Boal who coined the term "theatre of the oppressed" (see Boal 1985).

One of the practitioners interviewed for the Globalism Research Centre study was community theatre worker Rebecca Lister who said that when she finished a degree in social work at the University of Queensland she joined the Brisbane-based StreetArts Theatre Company that was consciously working with some of Boal's ideas.[5] When Rebecca's partner Tony Kelly got a job as a youth worker in the urban fringe suburb of Brisbane, Logan City, they decided to live and work in the same community and Rebecca got some help from StreetArts to start some theatre projects in Logan City, eventually forming her own theatre company called Feral Arts. When she was interviewed for the Globalism Research Centre study, Rebecca had moved and had been living in the

Victorian town of Daylesford for nine years where she had written and directed six ambitious theatre productions involving a very wide range of community participants. From the moment she first read Boal's work in the mid-1980s Rebecca "knew" that this was the work she wanted to do and she was delighted at the opportunities that had opened up for her in Daylesford; this also gave her a much deeper sense of belonging to her new community.

Given the legacy of people like Freire and Boal, many post-1970s community arts initiatives in Australia have had a political edge to them. For example, St Kilda-based visual artist and storyteller Julie Shiels began her career as a political poster maker in the early 1980s before teaching community groups how to make silkscreen posters. During the period of the radically conservative government headed by Jeff Kennett in Victoria she managed to sustain a visual arts program at the North Richmond Community Health Centre that was critical of Kennett's agenda and she said this taught her a lot about how to continue this kind of work in difficult political and funding environments. While they are working hard at their chosen art form and in building the necessary skills to work with a wide range of people, practitioners such as Julie and Rebecca have also had to acquire sophisticated project management and negotiation skills. It is no easy career path.

At the same time that many community arts projects and programs in Australia began to adopt a more overtly political stance, other community workers—including many trained in social work—adopted a "community development" paradigm that also emphasised the empowerment of disadvantaged people and communities by skilled community development practitioners (Ife 1995). While community development workers and community arts workers knew of each other's work, and collaborated occasionally, they tended to operate as separate domains until the early 1990s. Over the last decade, however, community development workers have probably gained a stronger appreciation for ways in which arts practices can give voice to the voiceless and make the invisible more visible, and community cultural workers have realised that they need community development skills to bring their art forms into community practice.

However, the Globalism Research Centre study showed that really effective community arts projects and programs need a combination of quite sophisticated community development and community arts skills and this can be hard to find in one person. Many of the practitioners in community arts projects and programs across the four communities in which the study was conducted[6] said that "tokenism"—i.e. addressing challenges in shallow ways—is dangerous when you start to engage with complex, and sometimes divisive, community issues. They stressed that the artistic elements of community art work need to be good quality, thus requiring people with good artistic skills. But when art workers challenge people to express themselves in new ways they need to be able to give them an emotional safety net or, at least, have other professionals they can call on if people become overwhelmed by what the experience evokes.

As indicated in the previous section of this chapter, the challenge for all community workers is to abandon limited, or even romanticised, notions of what constitutes community life in the contemporary world. The 1970s slogan calling on us all to "Think Globally and Act Locally" needs to be reworded as something much less catchy but more enlightening, such as "Think and Act in Local and Global, Real and Imagined Communities (all at the same time and with an adequate grasp of what the slogan actually means!)."

A time for reflection?

When Jon Hawkes wrote in 2001 of a need to integrate community cultural development work with public planning and governance he suggested that "cultural vitality" should become the "fourth pillar" of sustainability, alongside environmental responsibility, social equity, and economic viability. In opening the Fourth Pillar Conference in the Melbourne Town Hall three years later in 2004 he said that although there had been a proliferation of community cultural activities and supportive government statements and initiatives in the meantime he was "not happy" because most of the work now going under the rubric of community cultural development fails on the score of participation, authenticity, and vitality. He titled his paper "Let Them Eat Culture or Here Comes Another Bandwagon" and suggested that the new community cultural "industry" was being run by new industry professionals who have little or no commitment to the communities in which they work. At best they might provide fleeting opportunities for some people, at great economic cost to the community, and at worst they "scrape off scabs" without tending the underlying wounds, or compound the frustration of people who gain just a glimpse of something beyond the grind of difficult daily lives.

Hawkes made it clear that his critique was aimed at "opportunists" jumping on the "bandwagon" of cultural development rather than those who work hard to develop participatory community arts projects. However, some of the community arts practitioners interviewed for the Globalism Research Centre study felt that his sweeping criticisms had been unhelpful because they could undermine their efforts to promote community arts as a professional career choice. Such practitioners argue that opportunities for participating in art-making activities at a community level can be greatly enhanced by artists with "expertise" in the art forms being used, provided those artists also have an understanding of how to genuinely engage with people in communities. Too often, they argue, community arts have not really engaged people artistically and, at the same time, community engagement can be a valuable challenge for artists wanting to make their work socially relevant. However, it is not an easy career choice because, as mentioned earlier, an effective practitioner needs advanced skills in areas ranging from an art form to sophisticated project management to counselling. These practitioners felt that Hawkes had given too much weight to the dangers of opportunism and, in doing so, he had failed to acknowledge that many skilled practitioners are working hard to enrich local experiences of community arts with little public acknowledgement of the importance of such work.

The practitioners interviewed agreed that tokenism should be avoided and that funding bodies—including some local government authorities—impose requirements that make "authentic" art impossible. However, they feel that many practitioners have found creative ways to work with most of the constraints and avoid work that would compromise their own ethical standards. Many felt that opportunities to escape the tyranny of difficult lives, no matter how fleeting, can do more good than harm and some told of occasions when people came to them years after an event to say that it had changed their lives for the better (sometimes in totally unexpected ways).

Clearly it is to the advantage of people wanting to make careers in community arts to have enough resources to develop programs rather than one-off events, provided ongoing programs remain artistically fresh and stimulating for organisers and participants alike. However, arguments for funding must be accompanied by arguments for the need for such programs and arguments for the overall quality of community arts practices. Indeed, the Globalism Research Centre study suggests that the argument can be made that community-engaged art can sometimes be "cutting edge" in both content and expression because it works across a creative tension that many professional artists struggle to find. For example, no professional artist working alone could have conceived a single work replete with as many different stories as the Galgi-gnarrack Yiiranboi ("Weaving Lands") Tree that was created for the new Global Learning Centre in Broadmeadows by a multitude of weavers working from different weaving traditions under the artistic direction of Wendy Golden.[7] This project was conceived by the arts and culture director at Hume City Council, Anne Kershaw, and it was designed to highlight both the natural history and cultural diversity of Broadmeadows and to challenge some of the prevailing, negative, external perceptions of the local community. Similarly, public art works erected across the Port Phillip municipality in Melbourne under the Memories, Margins and Markers project reflect the capacity of artistic director Julie Shiels to work with local people in articulating their stories and finding an appropriate way to give expression to them "in place."[8] Only locals could have conceived of Hamilton's award-winning Top of the Town Ball that not only raised more than $250,000 for the local hospital but also gave all those who came to the elaborate event, held in large marquees on a site overlooking the town, a magical night to remember.[9]

It may be even more common in the field of community arts for works of art to become meaningful cultural expressions of complex lived experiences. Artistic expression can be one way of negotiating across the shared and divergent experiences of local community life in order to create a more tolerant and resilient society. The local can influence the global (or national) as well as the reverse. Perhaps this point could have been made more explicitly in David Throsby's valuable attempt to spark debate on whether or not we need a national cultural policy (2006).

Using research to enrich the field

A number of the community arts practitioners interviewed for the Globalism Research Centre study said that evaluation requirements for any grants received can be useful in making them think more deeply about what they are doing and how to make sure that their projects really do have the intended community-building outcomes. However, evaluation techniques are generally limited to feedback surveys that evoke quick responses from participants or audiences and observers. Such techniques can generate "snapshots" of a project in the language of statistics that can satisfy funding providers. But this is rather like taking still images of a fast-moving feast with a low-resolution camera. At best, the results might only suggest a need for more research.

For example, the Globalism Research Centre study included a survey of participants in a range of community arts events which revealed that people tend to participate at particular times in their lives, perhaps when they are new to an area or when they are spending more time at home. We included this question on the advice of project managers we had interviewed for case studies. However, the response to the question only suggests a need for further, longitudinal, research on the topic to know more precisely how and why people move in and out of community engagement.

Experienced community arts practitioners also told us that they would rather have opportunities to engage deeply with a small number of people than to have a more shallow impact on a bigger number of people, yet survey techniques tend to measure quantity rather than quality. Survey questions must be designed to evoke clear, quantifiable, responses so they are obliged to avoid nuances and leave out experiences that provoke more ambiguous responses. Rarely is there time or opportunity to encourage respondents to think carefully and reflectively about their responses, so a deeper processing of less conscious experience is hard to achieve with this technique. Survey techniques have their strengths but they work far better in combination with other research techniques that have different strengths (and weaknesses).

The Globalism Research Centre study used both a random survey of residents and a survey of people participating in community arts events, along with lengthy "strategic conversations" with practitioners, a collection of stories related to projects and activities across the four communities and the use of photo-narrative techniques designed to stimulate less conscious thought and reflection. It takes skill and experience to use each of these techniques well and time to use them in meaningful combination.

Social researchers have only recently developed a strong interest in "narrative inquiry"—i.e. an understanding of how and why people use stories to capture and communicate meaningful experiences. What they often lack, however, is a good grasp of the art of storytelling; the acquired knowledge of what it takes to either construct or identify stories that can move people in a range of ways. This is where social researchers have much to learn from arts practitioners who work with stories. In Victoria, for example, the best workshops on storytelling are probably those run by the writer Arnold Zable. Academic literature on the subject is, by contrast, rather dry and thin.

Researchers working on the Globalism Research Centre study also found that many arts practitioners and project managers are good at engaging in reflective conversations about their projects and experiences. Many of them noted that such lengthy "strategic conversations"[10] with sympathetic researchers helped them to reflect more deeply about their work and experience. This seemed to be a mutually beneficial form of exchange. We work on the assumption that local knowledge systems and expert knowledge systems can complement each other provided they interact with due respect for the distinctive character and role of each of these forms of knowledge. In this case, we can talk about the "insider" knowledge of community arts practitioners and the expert knowledge of social researchers being used in combination to create better processes for evaluation and advocacy, provided the practitioners and researchers both see a need for such a combination.

The integrated fields of community arts and community development have reached some limits in Australia and internationally because they are working with inadequate understandings of the nature of contemporary local community life. This is not the fault of practitioners because, as mentioned earlier, international literature that attempts to understand the global from the perspective of the local is disappointingly thin. We need more research that begins with the local and works towards the global (rather than the reverse), and this is already a theme of interest to community arts practitioners who are exploring the changing identities of local communities. Works of art often pick up emerging social trends long before other observers can detect them.

In his provocative talk at the Fourth Pillar Conference, mentioned earlier, Jon Hawkes warned that community cultural development work is in danger of becoming just another bandwagon for wagon-hopping opportunists who have little interest in authentic art or the capacity of art to express anger and frustration as much as joy. This fear seems overstated, at best, for there are no easy careers for artists wanting to work at a community level, especially those who favour participatory projects. However, tokenism is a serious threat to all forms of arts practice and for community arts practitioners this tokenism can relate to the quality of the artistic endeavour and the quality of the community engagement.

Community arts continue to be the forgotten sector of the arts "industry" in Australia but this neglect is now under challenge. A stronger partnership between community arts practitioners and social researchers seeking a more sophisticated understanding of the nature of community life in a globalising world can help to make a stronger case for greater public investment in this hitherto neglected branch of the arts.

References

Appadurai, Arjun. 1996. *Modernity at Large: Cultural Dimensions of Globalization*. Minneapolis: Minnesota Press.

Bauman, Zygmunt. 2000. *Liquid Modernity*. Cambridge: Polity Press.

Boal, Augusto. 1985. *Theatre of the Oppressed*. New York: Theatre Communications Group.

Brennan, Teresa. 2003. *Globalization and its Terrors: Daily Life in the West*. London: Routledge.

Falk, Richard. 1999. *Predatory Globalization: A Critique*. Cambridge: Polity Press.

Florida, Richard. 2002. *The Rise of the Creative Class: and how it's transforming work, leisure, community and eveyrday life*. New York: Basic Books.

—. 2005. *Cities and the Creative Class*. New York: Routledge.

Hawkes, Jon. 2001. The Fourth Pillar of Sustainability: Culture's essential role in public planning. Melbourne: Cultural Development Network.

—. 2005. *Let Them Eat Culture or Here Comes Another Bandwagon*. The Fourth Pillar of Sustainability Conference Papers. Melbourne: Cultural Development Network.

Ife, Jim. 1995. *Community Development: Creating community alternatives – vision, analysis and practice*. Melbourne: Longman.

Mills, Deborah, and Paul Brown. 2004. *Art and Wellbeing*. Sydney: Australia Council.

Mulligan, Martin, Kim Humphery, Paul James, Christopher Scanlon, Pia Smith and Nicky Welch. 2006. *Creating Community: Celebrations, Arts and Wellbeing Within and Across Local Communities*. Globalism Research Centre, RMIT University, Melbourne.

Nairn, Tom and Paul James. 2005. *Global Matrix: Nationalism, Globalism and State-Terrorism*. London: Pluto Press.

Savage, M., G. Bagnall and B. Longhurst. 2005. *Globalization and Belonging*. London: Sage Publications.

Steger, Manfred. 2005. *Globalism: Market Ideology Meets Terrorism,* 2nd edition.. London: Rowman and Littlefield.

Throsby, David and Virginia Hollister. 2003. *Don't Give Up Your Day Job: An Economic Study of Professional Artists in Australia.* Sydney: Australia Council.

Throsby, David 2006. *Does Australia Need a Cultural Policy?* Sydney: Currency House.

Young, Iris Marion. 1990. *Justice and the Politics of Difference.* Princeton: Princeton University Press.

—. 2000. *Inclusion and Democracy.* Oxford: Oxfòrd University Press.

Notes

[1] Transcripts of interviews with both Sulan and James can be found on the Globalism Institute website at www.communitysustainability.info.

[2] He made this clear in discussion of his essay on the need for a "cultural policy" (2006) at a public forum in Melbourne on 27 February, 2006.

[3] See www.communitysustainability.info for more details on this project and the methods used.

[4] The author speaks with the authority of one who has participated actively in many of the "social movements" that began in the early 1970s.

[5] The transcript of an interview with Rebecca Lister can be located on the Globalism Institute website www.communitysustainability.info.

[6] The Victorian communities centred on St Kilda, Broadmeadows, Daylesford, and the Hamilton region.

[7] See www.communitysustainability.info for interview transcripts and images associated with the community arts case studies included in the study.

[8] See www.communitysustainability.info.

[9] See www.communitysustainability.info.

[10] Typically they extended over a period of about two hours and required significant background research by the researchers so that they might hold up their end of the "conversation".

CHAPTER TWELVE

MEASURING THE SOCIAL CONTRIBUTION OF THE ARTS

EVA COX

Sharing experiences: undervalued aspects of the arts

A Sydney Festival event started my train of thought and stuck in my mind. It was *Sticky Bricks*, a performance by public housing tenants, in a long-term project with Big hART, telling their stories to the wider community and learning to like each other better through the process. The audience shared food, the performances and glimpses of the space and lives of the public housing tenants. The program stated that the group's key recommendation was to "make people welcome!"—so they welcomed us and made wider suggestions relating to inclusion and respect. This was a classic community arts project: a theatre group working with disadvantaged people to give them experience, confidence, links and capacities, and as a bonus, telling their stories to others in the wider world. This made me wonder about the artificial divide between these "types" of events and a visiting UK drama company's performance of *Tristan and Yseult* I saw in the same week. While there are many differences in production—creativity, experience, content, access and cost—there are also often undiscussed similarities in how such experiences may tie viewers/audience together by using the shared process to create new ways of seeing what others may see.

As a sociologist and researcher, I have been interested for some time in what makes us linked social beings rather than separate individuals. One option I have used for exploration is to look at social capital as a measure of our connections (Cox 1995). More recently I have been interested in the values base and quality of these connections, that is, what makes people more or less likely to "do the right thing" by others and to create a more civil society. The following chapter explores how some types of creative activities may contribute and create for all participants socially positive or undesirable outcomes and how these may be measured.

Critics like John Carey (2005) have challenged the concept of the civilising effects of the arts by pointing out how consumption and production of the arts have no necessary connection with more civilised behaviour. He notes that, along with language, conceptual thinking and other presumably exclusively "human" characteristics, consciously creative or decorative productions are part of all human societies. The modern world's placing of "arts" has often separated these activities from their social base, as is currently evident in Australia where the popularity and commercial valuing of indigenous arts ignore how the owner inhabitants integrated the creative with daily life and ceremony. The separation of performance and painting has occurred under modern influences, allowing these practices to be commodified rather then embedded in daily culture.

While expertise was presumably valued in indigenous society, it was more often the social purpose of the arts that was valued. This premise may offer a useful way of starting to consider how we can reclaim the social values of creative activities, aside from the financial or even aesthetic aspects. Reclaiming the more traditional initiatives of all present sharing the processes may offer models for revaluing socio-cultural experiences, which were, and still can be, part of what makes human societies both sociable and civil.

Questions of the aesthetic value of the arts are not my area. I am however the classic member of the audience who enjoys theatre, music and viewing other forms of creativity; this involves experiencing pleasures both in what I hear/see/feel and in most cases, sharing these experiences with others. As both a social researcher and as a person curious about what makes us social beings, I want to set a framework for defining and measuring how experiencing the arts may be part of the social connections that make human societies work, or not work.

My 1995 Boyer lectures, entitled "A Truly Civil Society," explored how we could tie social threads together in various ways to create the tensions and varieties that make for a vibrant creative society. I was concerned, even then, that we needed to recognise that differences and some conflicts were inevitable and also usually necessary to disrupt the harmonies that may make us too conformist. However, this recognition made it important that societies encourage the types of social connections that can both accommodate conflicting demands and interests and offer the resilience that allows such tensions to play out civilly.

These types of links can be defined in terms of "bonding" (intense), "bridging" (connecting differences) or "linking" (resources) social capital, the latter two being the links that tie the more intense groupings to the wider spheres. The mix is crucial as too much bonding and too little linking capital can play out as fragmenting violence. Debates on what makes us "social" have not been popular over most of the past two decades of neo-liberal and neo-conservative dominance, but more recently there has been a renewed interest in what makes us social. This still leaves the question of how one can measure these types of social links and validate their value to the broader society.

Finding the right measures

Many of the promoters of the arts since the eighties followed most other areas of public policy in seeking acceptance by applying economic language and concepts to fit into the neo-liberal dominance of policy, raising questions such as: Can the arts be appropriately described as an industry? This approach was used by many in the arts community to legitimate claims for public support and funding under neo-liberal policy constraints, made at a time when there was an almost totally economic agenda. Market forces were assumed to be the only determinant of success. Money was equated with value, so for instance the tax people doubted whether painters refusing to paint for the market were really professional. The effects of reducing such activities to their market value fails to value their contribution to identity, social cohesion, empathy, respect and other aspects of social wellbeing.

That was the eighties decade of greed. By the late nineties, there were signs that other ways of valuing were again being explored as part of wider debates on cultural connections and what makes us social beings. "Cultural capital" (often differentially distributed and usually individually owned) and "social capital" (group-based capacities to make connections with others and act collectively) were being discussed and defined to cover areas of daily life and social relationships that are not adequately represented by the trading model of markets. Like human and environmental capital, these terms were being added to audit and accountability exercises, but were often blocked when it was noted that there were no agreed measures for them. Developing such measures has proven to be a considerable challenge, as outlined further on, because the methodologies in use still tend to be looking for comparable data to those figures compiled by accountants for financial bottom lines.

The value of social measures of arts experiences

While there are still, in government and the community, many economic arguments about benefits and ownership in the arts "market," I am interested in looking at other aspects such as the role of experiencing and enjoying the arts through both public involvement in productions and being audiences. Sharing the experiences of creativity and its products often creates public, not private, good: ten people seeing a painting, or even a thousand, does not diminish the value of the painting or necessarily increase it, but can create a shared experience for strangers and offer points of implicit or explicit recognition of commonalties.

Joint ownership and access to group cultural activities can affect the way we see ourselves and others. Easy, widespread access to shared experiences can offer those involved symbols and processes that affirm belonging and democratic cultures. On the other hand, exclusion and ownership as symbols of power can reinforce the opposite. The occasional public struggles for ownership of certain artifacts indicates their potential symbolic power, like control over indigenous relics or the Parthenon Marbles. Like sport, the arts can often influence people's feelings and emotions, but we need to ensure that there is equal access so that these experiences do not become the province and toys of the rich alone.

One of the key questions raised by this position is the validity of using public money for universal access to certain experiences with the assumption that this is for the public good. The arts, at present at least, seem to bear some of the opprobrium of the term "elites": as an area where those who claim to be advocates for the commons can attack the high cultural spheres and collect populist plaudits. This echoes the Bourdieu (1984) research on art as another means of establishing distinction, of some tastes being "refined" as signs of superiority. The arguments often occurring around festivals and audiences may assume that there are certain values to be attributed to "high" arts. The public funding of the arts, in particular the proportions of funding that go to the major companies like the opera and state theatres, seems to explicitly suggest these offer better "value" because they serve elite audiences.

Although not the topic for this chapter, mentioning this debate is relevant because it explains both some of the defensiveness of arts practitioners and the basis for attacks on their funding. Assumptions about high culture as the province of the powerful fail to recognise the arrogance of such assumptions and the presence of appreciation and interests that often spread across the social spectrum. The question needs to be turned around to questions of whether there is public value in having cultural policies that democratise access to those areas assumed to be high arts, where expense may exclude many. Such policies and funding may give people a sense of sharing as citizenship; for example feeling welcome to access galleries and the sharing of experiences of music in free and affordable situations.

Who are the stakeholders and how is art being used?

The obvious participants in the arts are artists, performers, critics, curators, writers, galleries, festivals organisers, journalists, educators, students, buyers, collectors, viewers, audiences, employers, patrons, governments, tourists, and those involved in related businesses, spaces and media. However there are others in communities, such as particular identity groups, households, families and the wider public, that may want or need potential involvement to encourage their sense of being included. This raises questions on how the arts are being integrated into teaching and passed on through learning. How are the arts being used for celebrating, uniting or dividing communities, and as part of working, playing, living, belonging and excluding? What are its uses in differentiating, reconciliation and warfare? Not all uses are necessarily good and we need to recognise the possibilities of using art to express fears, to enhance the unknown, the transgressive, the mystic, and the mythic.

In times when a range of anxieties are almost encouraged by political leaders as justifications for increased social and political control measures, there may be good arguments for encouraging calm and social cohesion through diverse groups sharing these kinds of experiences. There are also other debates on whether using arts' expressions of anger and dissent may offer safer expressions of conflicting views than actual violence, as discussed below.

The arts as witness: sharing differing points of view

There are questions of the role of arts in expressing both official and controversial views. It is not surprising that officially approved viewpoints are promoted and alternative ones sometimes silenced, but in looking at social effects, we need to consider the role of those arts experiences that could encourage divisive, political viewpoints. Whether such material is appropriately defined as part of the arts depends on context and intention and maybe exclude that which is primarily intended to create violence and hate, like for example Hitler's use of symbols and spectacle. The arts can contribute to resilient social connections by illustrating different viewpoints and even bearing witness to aspects of society that may be seen by some as unjust or of concern. This type of process may be an important part of providing a platform for out groups and thereby encouraging civil dissent and more respect for differences. Given endemic diversity within societies, we need to recognise the often creative tensions of debates and encourage the types of dissent that may make societies self-aware and encourage change.

For example, in the recent past, plays, poetry, exhibitions and other forms of expression that relate to the situation of asylum seekers were important in making their case for more humane treatment. It is important to be able to recognise the ability of artistic expressions to confront, convert and create shared experiences amongst viewers that can contribute to respect for differences in viewpoints. There are many artists who see their role as part of social and political change activities. These types of activities can target people who can then see their own experiences and issues as part of the representations of the arts, as well as offering others some possibilities of sharing these experiences and the viewpoints of others. The ability to allow audiences to share such experiences in part creates trust and links.

Assessing the contribution of the arts through social wellbeing

The sections above illustrate some of the complexities of social connectivity that can be enhanced or undermined by particular types of participation in the arts. These experiential gains and losses for participants are not usually measured, as they require more sophisticated frameworks and methods than the easier body counts of more conventional "hard" data. The "soft" data on perceptions, feelings and reported responses requires using frameworks that focus on feelings, effects and links between people. This angle is reflected in discussions on measuring social capital which offers a useful starting point in exploring the concept of the arts as a social good. Social capital is useful because it is a collective measure, at least in the context I am proposing, in that it works as an indicator of the qualities of relationships that exist in organisations, groups and communities.

As mentioned earlier, social capital is now often defined in three ways: "bonding" social capital which ties us to intimates and familiars with thick ties, "bridging" which connects strangers and others with thinner ties, and "linking" which connects people to the institutions that distribute resources and govern us (Woolcock and Narayan 2000). And it is bridging social capital that is the core of this exercise. It covers those many less intense, often single-stranded relationships that lightly connect people individually, or groups sharing involvement, events, interests and spaces. This is often more important than our more intimate ties in looking at broad social cohesiveness and may also lead to linking social capital, i.e. access to formal resources and power.

Some of the material on social capital frameworks shows that "trust," of other people and of institutions, may be high in societies that are fairly closed, non-creative and monocultural (Putnam 2000; Leigh 2006). This statistic can be misused to argue for making societies that exclude troublemakers and those who are different, which limits creating social systems that can manage diversity and make societies fairer, with ethically diverse groups. There are some that promote the concept of value-neutral social capital but I am keen to complement such measures with further dimensions, such as valuing those measures that assess ways in which possible tensions can be civilly managed. This approach is particularly pertinent to arts practitioners because creativity is often a product of tensions and discomforts and is part of any cultural change. Adding values to social capital allows it to be used as part of social and cultural planning and activities, which may include many arts activities. This connection does not suggest a moral measure on the arts or any forms of control over artistic freedom but does provide another measurement for the value of public performances and events. This is a brief summary of a complex argument but raises issues about what kind of society we want to achieve and how we define and measure the necessary components.

The arts can offer useful activities and indicators of ways to increase cohesion in societies that recognise that in the current widespread mobile mixed populations, there are always going to be stresses and tensions. Therefore, arguments for funding should be able to include, where appropriate, encouraging activities that create social connections and mutual respect. If people can be encouraged to recognise both commonalities and accept differences, they create more resilient social connections. Effective social measures can use an ethics component with social capital measures to determine whether the produced social effects will be productive or detrimental to social wellbeing.

Using measures that assess the social value of participation in arts experiences and events seem to me to offer opportunities to measure viewer/audience connections with each other. Is there pleasure, and maybe a sense of future growth, in responses to arts experiences that make one feel good about not just self but others whom they share the event with? Even if the feeling is not necessarily fully communal or reciprocal, the possibility is that some types of shared feelings and experiences can contribute to the complex web that makes us feel more included and competently social.

To understand this we need to ask: can practitioners and/or audience involvement lead the participants to see themselves and others more positively? What forms of shared arts experience are pleasurable, exalting and inspirational? Which of these make us feel more connected and content, or more active and committed? When we see things that work on our passions, do they also make us feel more aware of others and our connections to them? Tallying the responses to such questions could become part of ways of counting the social contribution of the arts. Arts should not be pressured to make a social contribution but socially engaged art should also not be in any way devalued. Other arts experiences may not be overtly social but just sharing the pleasures should be counted where these experiences can both reinforce older and make new connections.

Setting the criteria

So how do we set the criteria? One possible dimension could be how the event itself encourages forms of sharing which are seen to be egalitarian and inclusive. This does not mean that everything has to do this but where this is intended, the experience offered contributes to perception and sense of belonging without demeaning or distressing some participants. There are also questions of who fails to gain access because of institutional barriers and costs. For instance, what is the social difference between opera at the Sydney Opera House and Opera in the Park with 100,000 others? Can wide access to an ordinarily expensive elitist mode of music bind diverse groups via *Tosca*?

This is one aspect of inclusivity but there are other more complex questions if one is not to limit shared experiences only to those that are uncontroversial. How effectively can the arts raise questions about the richness of diversity, tastes and experiences and assist in making connections that over-ride despite manifest unfairness and injustices? Do these varieties of viewpoints allow audiences to share unfamiliar experiences and make connections, which could be assessed in terms of their social value? If the arts in these roles provide legitimation of those groups who are judged as outsiders, they may create the will for resolving conflict civilly. On a lesser level, the arts may allow people to share the experiences of strangers in ways that make them form useful social connections that would otherwise be absent

Making measures

The questions of how we measure the results of arts and cultural activities come back to some of the basic debates on the measurement of human viewpoints. There are many economists and public servants who demand statistical, big number measures to prove the worth of their funding. Such data may be both expensive and not very accurate in really representing complex viewpoints. Lots of people can signal agreement to simple questions and offer some figures to keep the bureaucrats happy with superficial items. Good research needs to mix methods such as observation, in-depth interviews, focus groups and even some forms of before and after measures to more accurately record and explore changes in viewpoint.

How to measure which experiences of the arts make the participants and creators more able to connect or more resilient in their social connections may require more work, including eclectic experimentation with methodologies and metrics. We need to be able to find measures that will satisfy the need for in-depth understanding of audiences and participants' feelings and attitudes as well as create easy counts that satisfy funding bodies.

How does the concept of a public good fit into the area of the arts? Looking at participation would involve much more than just counting bums on seats. Part of the basic defining of social capital puts a strong value on organisations that involve people and on participatory types of learning. Therefore measures are needed of participants' feelings and reactions; for example, their enjoyment of the experience, their feelings of acceptance and pleasure in the "show" are all contributing factors to be taken into the value of their experiences.

I have deliberately not looked at specifics of measuring the effects of events like concerts, art shows, performance, theatre etc. because they can be examined in many different ways, apart from just counting attendances. There are many ways of categorising effects that are related to outcomes and can be tailored to fit a range of experiences. I am here arguing for a new approach by looking for the possible extra social dimensions of the event under study. The emphasis should be on how the events affect diverse participants such as the experiences of maybe working together or just being able to experience sharing the viewing of the work with others. This approach assumes that the group process of viewers and audience is a major part of the experience, which is not currently measured.

The following are some examples of possible measurable activities that can offer starting points for devising questions to assess the social value of arts experiences.

- Collaborating on some creative process/product with others
- Experiencing shared involvement in creativity with strangers
- Sharing audience/viewing spaces with strangers
- Having access to non-majority cultures in public places
- Seeing others appreciate creative versions of your cultural experiences
- Recognition of diverse heritages: one's own and others'
- Being able to experience the viewpoints of others
- Seeing works that offer alternatives to the received versions of dominant groups
- Bearing witness, sharing experiences others have had
- Exploring joint experiences with others
- Sharing pleasure and excitement with strangers
- Experiencing your own and others' reactions to examples of excellent creative endeavours
- "Being there" i.e. being present when something is happening and being able to define yourself as being part of it.

The following offer some examples of the categories of types of questions that could be used in assessing the social value of such experiences:

Being there experiences

- Who do they go with? (Did they take friends, family children?)
- Who did they see there?
- Did they meet friends?
- Did they talk to strangers?
- What other sorts of people were there?
- Were there connections established with others who attended?

Feelings about event

- How did people feel about being there?
- Did they feel they belonged?
- Were they welcome and comfortable?
- If not, why did they feel unwelcome?
- Were there others that were not welcome?
- Were there groups who would not have felt welcome there? Who?
- How did they feel about the production/exhibition?
- What did they remember seeing?

Did anything make them:

- feel good?
- reinforce their sense of who they were?
- feel bad?
- challenge what they thought?
- understand other viewpoints?
- feel they wanted to know more, to follow up?

Reflecting on the event later

- Did they tell others about their experiences?
- Would they go again? Why? Why not?
- Who else do they think should go? Would go?
- Who shouldn't go? Is there anyone who wouldn't like it?
- Would they go again?

Longer-term effects

- What benefits were there from such performances/ experiences?
- Can it make people feel differently about particular related

issues or other groups involved?
- Does it result in future involvement in community/creative events and issues?

There are some available scales on trust to see whether audiences' experiences have resulted in lower or higher levels of trust but these would need to be developed for this type of application.

The research could also include sets of questions on the effects of being involved in the processes of creating performance or other productions. Questions for participants could explore the social aspects of their involvement, both in relation to the process and to gauge their expectations of audience reactions. Questions for the artists could include:

- Are there social aspects to why they create?
- What do they gain from creating, in terms of changes to their perceptions and relationships?
- What responses do they want from the audience/viewers?
- Why do they want that?
- What effect do such responses have on them?

The above material is a tentative exploration of the dimensions of social measurement of arts experiences. There are many other aspects that could feed into social measures that have not been included. My recent experiences of participating as a member of the audience at Opera in the Park reinforces my perceptions that people develop goodwill in these events and are generally very courteous, pick up much of the rubbish and are patient as the big crowds move off. However, I have not seen any measures of these effects. There are other techniques that should be employed, including formal observation of what people do and how they act and react. This recording of what people do is important because people are often poor reporters of what they have done and how they interact; so observation adds another dimension.

Existing initiatives

The following items, lifted from a local government cultural planning document (Lee Shoy 2004), show how some of the above areas are already partly on the agenda. In Sydney, the Western suburban councils are looking to the arts for:

- community identity and cohesion
- cultural integrity—continuation of local culture, tradition, rites
- peoples' histories
- connection to place, belonging
- perceptions of community cohesion
- cultural inequities.

Involvements in local arts activities are expected to lead to:

- intergenerational contacts
- increased use of city centres
- reduction in fear of crime
- support for vulnerable groups
- participation in local consultations.

Conclusion

We need to understand more about why people want to be involved with others and/or watch what others are creating and what social benefits accrue by taking part. If people are part of an audience at a concert/play or other event, what do they gain from being there? We also need to recognise that sometimes such activities can create negative and anti-social reactions as well as pro-social ones. One of the proclaimed virtues of wider access to different versions of heritages and histories is the sense of shared pasts, of connecting with others because of the communality of perceived experience.

There are some socially oriented uses of the arts that I am not including in the scope of this chapter, as their primary purposes are very different to the creative enterprise. An example might be official propaganda employing negative reinforcement of dominant views or that "showcases the exotic" in ways that deliberately demean or objectify groups. These uses raise other issues not canvassed here. There are also differences when the purpose of creative endeavour is primarily therapeutic, such as the use of graffiti projects, filmmaking, theatre, music and art as part of treatment programs. Neither propaganda nor therapy should be confused with the social assessment of the general run of exhibitions, events and performances, as their rationale engenders different expectations to those that general arts events seek to achieve. That is not to undermine the importance of the arts as therapy, but to suggest that its measures would need specific designing.

With the above negatives and caveats in mind it would be useful to do more work on defining and measuring the general social value of the arts. Too much of the substantial public funds tied into both funding arts production and its display and distribution is based on its economic value or cultural aesthetic. If we are to broaden the appreciation of the arts into another area, we need to add a third bottom line: the social/ethical measure of its values. Without saying all arts would merit such measures, it would be good to include those that do. We need to identify which types of events, endeavours and productions could be seen as offering ethical and social gains. These measures could then be offered as further evidence of the value of certain types of arts experiences to social wellbeing for all. The question is whether the results create more sociability, more comfort with strangers and greater competence in making connections.

References

Bordieu, Pierre. 1984. *Distinctions: A Social Critique of the Judgment of Taste*. Cambridge: Cambridge University Press.

Carey, John. 2005. *What good are the arts?* London: Faber & Faber.

Cox, Eva. 1995. *A Truly Civil Society*. Sydney: ABC Books.

Leigh, Andrew. 2006. Trust, Inequality, and Ethnic Heterogeneity. Discussion Paper No. 511, The Australian National University Centre for Economic Policy Research

Putnam, Robert. 2000. *Bowling Alone*. New York: Simon & Schuster.

Throsby, David. 2005. *Cultural heritage as financial asset in strategies for urban development and poverty alleviation*. International Conference for Integrating Urban Knowledge & Practice, Gothenburg, Sweden, May 29–June 3.

—. 2003. Economics and Culture. *The Quarterly Journal of Austrian Economics*, 6, 2 :103–106

Tiffany, Lee Shoy. 2004. Cultural Indicators–an exploration with Colin & Sharon (unpublished paper).

Woolcock, M. and D. Narayan. 2000. Social capital: Implications for Development, Theory, Research, Policy. *The World Bank Research Observer*, 15: 225-251.

CHAPTER THIRTEEN

CITIZEN VALUES AND THE ARTS: NEW LANGUAGE FOR AUSTRALIAN CULTURAL POLICY

DEBORAH MILLS

Introduction

Areas of public policy only become possible when they become thinkable.

Is the idea of an Australian cultural policy thinkable? I guess the answer to this depends on what kind of policy you are after.[1]

I like to think that a cultural policy should illuminate what it means to be human and not just cover the role of "art in government," or at least not only this. An argument on the economic value of the arts—begun by arts bureaucrats in the 1980s as one means for justifying government expenditure on the arts to politicians, who, if not actively hostile, were at least indifferent to government expenditure in this area—has become *the* rationale for the role of the arts in government. As a result most politicians, public intellectuals, commentators and arts bureaucrats appear to have lost the language necessary to express the intrinsic importance of art and culture in our lives.

A reason for pessimism about the development of a national cultural policy is the simple fact that making and enacting cultural policy is difficult (and I examine some missed opportunities and suggest the reasons for this later in this chapter).

Another factor militating against the development of a national cultural policy is that much of what passes for arts and cultural policy in Australia fails to clearly articulate the "values" upon which it is based. Where the policy values are implicit they become invisible and difficult to argue for or against until the consequences are made apparent through action or inaction.

Yet the evolution of the policy and practice of community cultural development has been fraught with overt and covert struggles over values. These struggles have been with community cultural development practitioners and practitioners of "mainstream" arts, and also between different practitioners within the community cultural development sectors.

It is for these reasons that I think that a study of community cultural development policy and practice in Australia can illuminate some of the dilemmas to be confronted in developing a national cultural policy. In this context I will explore the tensions between policies aimed at improving "access" to the arts and those to improve "participation" in arts activities.

A problem of definition

Within what has been referred to as "community arts" or "community cultural development" there has been a diversity of practice that has made it difficult and somewhat contentious to define. The terms and definitions we have are, in fact, bureaucratic in origin (Hawkins 1993, xviii), and over the last thirty years the practice has been variously defined by the Community Cultural Development Board of the Australia Council—and its predecessor, the Community Arts Board—and practitioners as:

- The teaching of arts skills by professional artists. For instance, a professional writer or visual artist teaching people how to write or draw.

- The development of work by professional artists that is informed by the experiences of members of a community. These works often tell the stories of members of that community and may involve, for example, the development of a play by professional writers and actors with members of the community acting as dramaturgs. Sometimes these works may have a social change objective.

- The active participation of members of a community in collaborative creative practice. For example, the development of a play or public artwork where members of a community collaborate in writing and acting, or designing and manufacturing. This approach usually involves professional artists working alongside members of a community imparting skills and offering advice. The term "collaborative practice" refers to the exchanges of knowledge and skills which take place between participants.[2] Sometimes these works of the imagination are also "cultural activism"—aiming to raise awareness about an issue and mobilise social action.

These definitions, and others, have been contested by people engaged in this work and by the Board itself from time to time, and the controversy can be better understood by focussing on the *intent* of the practice.

For some practitioners and policy makers these practices are about "celebration," "civic enhancement" and/or "skills development," and are arts-based, of (usually) limited life, and "make people feel better."

For others the practice is about "collaboration"—people working together using a wide range of creative processes (not exclusively arts-based) to illuminate aspects of their lives and to bring about change. Here the emphasis is on a process of community empowerment over time.

Whatever the intent, discussion about community arts and community cultural development has tended to emphasise the "ideas" and the "process" of that activity, rather than the art form used or the art work emerging from it. This emphasis was given impetus when the Community Arts Board rejected "art" in favour of "culture" in 1986, defining culture as:

> the way we pass on, experience and explore our values and views… communicate with each other, interact with our environment and organise ourselves (Hawkins 1993,157).

The establishment of the Community Cultural Development Committee[3] at the Australia Council in 1987 was both a clever action by the Council's Chair, Donald Horne, to preserve the "old order" of the Community Arts Board in the face of attempts by the then (Labor) Arts Minister to abolish it, and an attempt to entrench the broader "cultural" scope of the Board and to move beyond a "grants culture" through building strategic alliances with other institutions (Horne 2000, 272).

In 2006, at the instigation of the Australia Council, the Community Cultural Development Board was abolished by the (Liberal) Minister for the Arts and replaced by a new section, Community Partnerships, located in the Audience and Marketing Development Division and answerable to the Council rather than to a Board. The action provoked outraged response from many in the sector (myself included). Apart from the diminution in status, by locating "cultural activism" within "the development of audiences for arts product" the Council's act struck at the core values underpinning community cultural development policy and practice.

The Working Party established by the Australia Council in response to the public outcry was asked to develop a policy framework, goals and objectives for supporting art "in and with communities" (Dunn 2006, 40). In their *Report of the Community Partnerships Scoping Study*, a definition of what is meant by art "in" and "with" communities was sidestepped in favour of a recommendation that the Council "broaden the definition of community arts and community cultural development to include the wide range of arts and cultural activities currently practiced in communities" (Dunn 2006, 8). Whether this sidestepping results in a de-politicisation of community art and community cultural development by the Australia Council remains to be seen.

Art does not equal culture

When Donald Horne and Jon Hawkes came up with the idea of a Community Cultural Development Committee they intended to broaden its work beyond what they felt were the "too narrow" definitions and conventions of "the arts" (Horne 2000, 262).

Other cultural policy makers and administrators have been less clear in their intent and often use the terms "art" and "culture" synonymously. Possibly they think culture has the broader appeal and helps bring the arts in "from the margins" of government concern, or perhaps they are insisting on a distinction between "prestige" art and "community" art. That they are often unable to follow through with any sensible discussion about culture only further, it seems to me, entrenches the arts in the margins (Mills 2001).

Culture and its role in everyday life are not widely understood in government. Where governments do talk about "culture" in the title and introductory pages of their policy documents they quickly retreat to "art" when it comes to priorities and strategies. For example, in the Queensland Government's 2000 *Creative Government* policy the arts is used synonymously with culture and applied as an instrument of government policy, and in Bob Carr's NSW Government's cultural development policy from 1999, the terms "art" and "culture" appear to be used interchangeably (except when used to distinguish prestige art from cultural activity at the community level).

To some extent this also the case in the Federal ALP arts policy which is resolute in its use of the term "art" except when referring to activity in local government and regional areas when the term "culture" is preferred (Garrett 2007, 7).

By way of contrast, the *National Culture/Leisure Industry Statistical Framework* endorsed by the national Cultural Minister's Council went well beyond "the arts" to include: the structures and skills for the management and development, distribution and marketing for the products of the arts, crafts and cultural industries; commercial and public cultural facilities; the natural and built environment; local traditions; Indigenous culture; ethnic and cultural diversity; and heritage (Cultural Ministers' Council 1991).

In practice, whatever the policy conception of "culture," the application of cultural policy is too often reduced to a heritage and subsidised arts scope.

To develop a national cultural policy, governments at all levels will need to develop greater sensitivity to and an awareness and appreciation of our culture and what it means in everyday life, then apply this understanding to the ways in which they plan for our society and our communities. This application needs to occur in a transformative way that moves beyond the instrumental and allows us to imagine new possibilities.

Instrumental and transformational approaches to policy

This idea of the transformative nature of art informed the writing of *Art and Wellbeing* (Mills and Brown 2004). Paul Brown and I were commissioned by the Australia Council to identify where and how community arts processes enhance the efforts of government agencies concerned with community and individual wellbeing.

For the project we adopted a concept of "wellbeing" which builds on a social and environmental view of health and that recognises the inter-relatedness of social, cultural, economic and environmental factors.

Our research found that the understanding and application of the arts in the work of government has focussed, for the most part, on applying the arts in an instrumental way. By this we mean, "let's implement policy using the arts as a tool." In this sense the arts have served, for example, as a focus for promoting a city, as a way of educating communities about environmental issues, or as a tool for civic enhancement.

However, we propose that there are transformational possibilities arising from engagement with the world of the intellect and the imagination which move beyond these instrumental applications. By "transformational" we mean, "let's allow creative activity to help determine policy by developing and negotiating shared understandings of various policy challenges and mapping out solutions."

We argue that creative processes and our critical engagement with the material culture these processes create—that is, art—can free us from the traps of habit, help us to see things from a different perspective, suggest connections between varied subjects and transform communities and the way in which government operates.

Transformational approaches can unlock new solutions to the challenges faced by all governments working for the wellbeing of their communities. However, it will only happen if governments develop their "cultural sensibilities"—learn how to recognise the cultural dimensions of what they do, become more self conscious about them, if you like. They must also recognise how habits of thinking, seeing and behaving can be transformed through active participation in creative processes.

The difficulty those working in the arts have in understanding and applying this transformational approach is illustrated in the Foreword to our book *Art and Wellbeing*. Written by then CEO of the Australia Council, Jennifer Bott, the book's publication is justified in terms of the efficacy of the arts in strengthening social capital—despite this being directly at odds with the main thrust of the book's argument (Mills and Brown 2004, 2).

More recently the *Report of the Community Partnerships Scoping*

Study, the rationale for government investing in community arts and culture concludes that they are:

> a vital link between the arts and cultural sector and the everyday lives of the broader community, and deliver a wide range of arts, cultural, social and economic benefits. (Dunn 2006, 3)

These examples suggest that instrumentalism is well and truly entrenched, at least in the Australia Council's thinking.

The fundamental problem with instrumentalism is a failure of imagination, a refusal to acknowledge the new policy opportunities or the new realities that art and cultural development processes can open up. Australian governments have too often failed to recognise or engage with their own substantial influence on culture; an influence far greater than publicly subsidised arts activities. For example, land use planning policies that approve the construction of shopping malls have cultural consequences, including the reduction in public places and spaces, and the demise of small cafes and corner shops where neighbourhood stories are exchanged, local notices pinned up and social and cultural capital developed (Mills 2003).

A technocratic or values approach to policy development?

Another feature of the "art equals culture" misnomer is that the policy proposals usually ignore "mass" and "popular" culture entirely. Why, when their impacts so obviously outweigh those of traditionally subsidised arts activities? To leave them out appears to be irrational.

This brings me to my second theme, namely the irrationality of public policy making. [4]

Policy making by government is usually opportunist and is *really* informed—sometimes unconsciously—by values rather than evidence. The role of values in policy is made particularly clear from observing the different ways in which governments in Australia have treated art and mass culture. Why do governments have different organisational relationships for the support of art and mass cultural product? Why do they make different assumptions about the product, its distribution and the audience potential?

While it can be argued that these distinctions are based on assumptions about market failure, I argue that they are also based on making moral distinctions between different forms of creative production. For example, the Australia Council privileges certain forms of cultural production over others because it sees them not just as different, but as better and worthier than other forms. The Council appears to truly believe that with some effort at education and, more recently, marketing, subsidised arts activities will be universally appreciated and enjoyed.

Federally "the arts" are seen as separate from "media and communications," coming under different government departments. With the Department of Environment, Water, Heritage and the Arts now responsible for the policy and content of new media and the Department of Broadband, Communications and the Digital Economy responsible for technological development. The arts are seen as "non industrial" and production is subsidised while the media are viewed as "industrial" where the role of government is to regulate their distribution and exhibition. The exception is film-making which the government treats as art and subsidises production, not distribution, ascribing to it the high moral value of contributing to our national identity (Gonski 1997, 15).

The reasons for this are not "rational"; they are based on the collection of beliefs and values underpinning and driving the policy "moment." Unless we can unpack, analyse and interrogate these values then we will not construct internally consistent cultural policy.

The value of the arts as a "civilising influence"

Utilitarianism and a belief in the civilising influences of the arts and culture underpinned both the development of Australia's public cultural institutions during the nineteenth century and the establishment of the Community Arts Board of the Australia Council during the latter half of the twentieth (Hawkins 1993, xviii). These beliefs can still be seen in the policies of the Australian Liberal and Labor parties and the Australia Council.

Far from being egalitarian, these policies are grounded in a belief in the distinction between those who appreciate art and culture and those who do not, and in the need to address this "cultural deficit" through remedial action such as funding for touring or audience education programs which will attract people from less "enlightened" socio-demographic groups.

Government policies on the arts may purport to be egalitarian but the wealth of evidence shows that audiences for "the arts" are tertiary educated, middle-income groups. What discernable impact has this evidence had on arts expenditure? Little. Do arts bureaucracies seek out other, more popular kinds of art to support? Not to any significant extent.

It seems bureaucrats and politicians still believe that there are two distinct categories of art and culture; those for the elite—"art"—and those for the masses—"entertainment"—and that mass audiences cannot appreciate high art (Bennett 1994a, 17-23).

Whereas postmodernism may have challenged the old vocabularies and hierarchies of distinction between high and low culture there has been no evidence that these views are effectively challenging the dominant values of the policy makers. When Donald Horne proposed, as the Australia Council's Chair, that one of the three cultural rights was "the rights of citizens to engage in their own forms of intellectual and art production" this "was scorned as unprofessional by some of the Council staff, by a few ut the successes of CCD□ XE "community cultural development" □ as an approach which goes beyondt strongly" (Horne 2000, 271).

However, the arts policies of the former Howard Liberal Government and current Labor Government do reveal an enthusiasm for the Australian people making their own art (albeit under the guidance of professional artists). While the Howard Government's policy placed great emphasis on a "cultural deficit" approach and supported touring visual and performing arts to rural and regional Australia—citing the "significant cultural, social and economic benefits" (Liberal Party of Australia 1996, 8)—it also supported the Regional Art Fund as enabling "regional communities to generate their own art" (Liberal Party of Australia 1996, 10). The current Labor Government's policy has moved a long way from the Labor Party's earlier policy emphasis on "professional" arts (Australian Labor Party 2005, clause 11) to acknowledge the importance of the creative expression of all citizens. Whilst recognising that the arts are "worthy of support in their own right" the policy also values the economic, health, educational and social benefits of the arts (Garrett 2007, 1).

Access versus participation: the great divide?

The Australia Council was established in 1975 to support excellence in the arts and to ensure that all Australians have the opportunity to engage with the arts and enjoy a rich cultural life (Commonwealth of Australia 1975, Section 5).

How Australian's "engage" has been interpreted by the Council, and other arts agencies, as encompassing both access to the arts and participation in them. During the last twenty years there has been a clear distinction between government policies and the activities of arts organisations which aim to give Australians opportunities to critically "engage through having access" to their cultural heritage and to new art, and those polices and activities which aim to provide opportunities for individuals and communities to "participate" in their own forms of intellectual and art production. On face value these policies and activities do not appear as polar opposites. However the values underpinning them are very different.

The values of *engagement* are associated with conventional "high" culture and the desire to give people better access to it. The values of *participation* are associated with valuing a diversity of cultures and an understanding that popular and marginalised cultures can invigorate high culture.

The values of *engagement* are associated with a singular view of citizenship and nationhood, what Mary Kalantzis and Bill Cope refer to as an active homogenisation of national interest (1995, 51). The values of *participation* are associated with diversity rather than singularity and a belief that people can belong to different cultures simultaneously.

The values of *engagement* are associated with an elite of institutionally recognised artists. The values of *participation* are associated with many different definitions and interpretations of virtuosity.

The values of *engagement* support a singular, dominant cultural heritage. The values of *participation* see many cultural heritages conversing with each other.

The values of *engagement* value the negotiation that takes place between artist and audience around that engagement. The values of *participation* value the capacity of communities to be active participants in the making and management of their creative and cultural development.[5]

These two points of departure; "engagement through access" and "participation" began in the same place thirty years ago in the work of the Community Arts Board. "Champions of Engagement" have arisen; for example the Arts Councils in each State and Territory which were established to tour music, dance and theatre originating in Australian capital cities to rural and regional areas. "Champions of Participation" have also emerged; for example, some of the peak bodies known collectively as "Networks" established by the Community Arts Board to provide information, training and support to the community cultural development sectors. Until the recent past each had hardened their approaches into dogma which each view as oppositional to the other.

This dogmatism has had at least two discernible effects. The first or these has been in the dichotomy which has arisen between each of our three "cultural rights," as Donald Horne would describe them, namely the right of all citizens to critically engage with their cultural heritage and with new intellectual and artistic production and to find their own forms of intellectual and artistic expression (Horne 2000, 252).

Champions of Engagement have tended to focus on the rights of citizens to engage with their cultural heritage and somewhat overlooked the rights to engagement with new intellectual and artistic production and to finding their own forms of intellectual and artistic expression. Champions of Participation have tended to focus on the rights of citizens to find their own forms of intellectual and artistic expression and only in a narrow ethnocentric sense to have access to their cultural heritage, while the rights of citizens to critically engage with new forms of artistic and intellectual production has been somewhat overlooked.

The second effect, which could be said to be a natural consequence of the first, is that policy makers have been allowed to ignore the fact that despite the high level of engagement and participation by Australian citizens in a range of popular, "sub-cultural" and "high" arts activity, the vast majority of them do not have an opportunity to engage with the output of subsidised arts and cultural institutions.

More recently, some major arts institutions, like the Arts Councils, have ignored this dogmatic divide and, to some extent, their own history, and made significant steps towards acknowledging the need to engage with a diversity of cultural heritages and to support opportunities for communities to participate in their own forms of intellectual and art production (Regional Arts Australia 2005, 9).

The report of the Australia Council's Scoping Study indicates a similarly pluralist approach,[6] recommending that the Australia Council adopt the following goals for its support for art in and with communities (which it subsequently did):

- To create opportunities for communities to engage in the exploration and development of heritage, culture and contemporary art.
- To encourage individuals and communities to share and participate in their own forms of arts, cultural and creative expression.
- To value and encourage the arts, culture and creative aspirations of communities.
- To support and promote the multiplicity of experiences, interests and backgrounds underpinning Australian communities (Dunn 2006, 19).

While falling short of expressing these goals as cultural rights, these sentiments are not far removed from the Citizen's Charter of Cultural Rights in the preamble of the *Creative Nation* statement, the first and last attempt by an Australian Government to develop a cultural policy (Commonwealth of Australia 1993, 6). Written in the last days of the Keating (Labor) Government the policy was abandoned by its successor, the Howard (Liberal) Government.

"now you almost see it, now you don't": Government approaches to cultural policy

Do our ideas about the role of government include support for cultural development?

I don't think so. There has been a squeamishness with which Australian governments of all persuasions approach the issue of the involvement of the state as a significant or dominant force for cultural development. There is a double standard with the mechanisms of arms-length funding and peer group review used as a shield against making explicit the values that underpin current policies and decision-making processes. We have had invisible cultural policy—only made visible in the actions or inactions of governments and their agencies—and the field has been left wide open for Ministerial bureaucrats and politicians to exploit this invisibility according to political whim.

The former Howard Government overcame their squeamishness to fully exploit this vacuum and expand the range of ministerially controlled funds while attempting to suppress artistic content if it smacked of dissent.

In his 2005 Philip Parsons Memorial Lecture, Australian author and journalist David Marr cited several examples of that Government's interventions to prevent theatre from being seen which might be interpreted as critical of their policies or their politicians. *Through the Wire* written by Ros Horin, then artistic director of Sydney's Griffin Theatre, was denied funding to tour from the Arts Minister's Playing Australia Fund. As Marr said:

> What appears to have happened at the meeting of Playing Australia last year was this: despite the show having a very high score on application, the Minister's representative persuaded the committee not to recommend it for funding on the basis that it was not yet a fully fledged production. Other shows were rejected at the same meeting on the same unexpected grounds… In the industry there's little doubt that Canberra was simply not going to back a politically unpalatable show (Marr 2005).

He also described the rejection of funding to tour a new work about the Iraq War, *The Wages of Spin,* which had received Australia Council's Theatre Board's support; with the reason given it was "too capital city focussed" (Marr 2005).

Marr also related an incident where then Federal Arts Minister, Rod Kemp, was sufficiently moved by the Melbourne Theatre Company's production of Hannie Rayson's play on the subject of boatpeople, *Two Brothers*, to ask to that Theatre Company's Chair, "Why do you persist in biting the hand that feeds?" (Marr 2005).

Such intervention was not restricted to the performing arts. The Howard Government's attempts to control the exhibitions and programming of the National Museum are described by David Throsby as "more interventionist than any Labor government has been" (Throsby 2006, 10). When the now defunct New Media Board of the Australia Council funded five video artists in 2003 to make a video game called *Escape from Woomera* it drew strident criticism from the Minister for Immigration, Philip Ruddock, *and* Arts Minister Kemp.[7]

The current Labor Government policy asserts a commitment to restore funding decisions to the Australia Council and to strengthen its peer review and arms-length funding processes (Garrett 2007, 4). It is too soon to say whether and how this will happen.

However, the policy appears to be imbued with the same old values, making only a fleeting reference to "access and participation" in the arts (Garrett 2007, 7).[8] And the hierarchies of distinction between high and low culture are not tackled head on except in terms of industry development for contemporary music and new media (Garrett 2007, 9, 10 & 14).

Some missed opportunities for cultural policy

Even if a national cultural policy were imagined, a government needs to be willing to act on it and the organisations who deliver the policy must understand and agree with it. There have been some missed opportunities in cultural policy in Australia due both to a lack of political will and the failure of bureaucrats to either comprehend or, due to a clash of values, cooperate.

In Western Australia in 2002 the work of the Sustainability Unit on the *Focus on the Future* strategy presented an opportunity to illuminate the cultural dimensions of economic, social and environmental planning and their interdependency (Western Australian Government 2002). However, this opportunity was lost due to the inability of anyone in government, or close to it, to understand or articulate the cultural dimensions of these other aspects of government policy. The department responsible for the arts, instead, advanced an argument for greater funding for subsidised arts activities as its contribution to this policy initiative.

In 1995, Colin Mercer, Lesley Fogarty and Jonah Jones developed a Cultural Policy Framework for the now defunct Aboriginal and Torres Strait Islander Commission (ATSIC). Their framework was designed to function *simultaneously* as a cultural development and economic development policy and it embraced the full range of cultural development practice, including media, cultural heritage and popular culture.

Two years later this work was subsumed by ATSIC's Cultural Industry Strategy. This strategy was a far cry from the original framework, reverting to "colonial" definitions of the arts—such as dance, music and crafts—and ignoring the significance of cultural issues and practices such as relationship to the land, a sense of place and the role of traditional foods and medicines. The later document also suffered from viewing cultural industry strategy as the economic arm of a cultural policy rather than as a manifestation of the policy itself.

In 1991 Joan Kirner's Victorian State (Labor) Government produced a cultural policy called *Mapping our Culture* that attempted to enshrine the cultural rights of individuals and communities. While it could not be expected to survive the Kennett (Liberal) Government, sadly, the Bracks-Brumby (Labor) government overlooked this extraordinary policy in favour of creating one of its own.

A government's choice of organisation through which to deliver its policy is also significant. For many years the policies of the Community Cultural Development Board sat uncomfortably within the Australia Council. As far back as 1985 the Board's *Programme of Assistance* booklet defined art as a "subset of culture" and articulated a set of cultural policy values drawing on an anthropological, rather than an aesthetic, frame of reference (Hawkins 1993, 157). This policy challenged many of the deeply embedded, and often implicit, values of the Council including:

- *Aestheticism:* particularly who has the power and authority to create and to decide what is beautiful. The Board challenged the canon of institutionally recognised artists and accepted many different definitions and interpretations of virtuosity. While the arts were acknowledged as "the paramount symbolic language through which shifting meanings are presented" (Hawkes 2001, 23), a broader definition of creative endeavour was acknowledged and supported. The Board also understood that popular and marginalised cultures can and do invigorate "high" culture and that Australia has a diversity of cultural traditions and that people can belong to different cultures simultaneously.

- *Utilitarianism:* the Board did not subscribe to the belief in the distinction between those who appreciate art and culture and those who do not, or to the need to address this "cultural deficit" through remedial action.

- *The artist as hero:* the Board's belief in the creative capacity of groups and communities also challenged the hegemony of the belief in the individual artist as sole creator.

Defining a new set of principles for a national cultural policy

The symbolic power of policy is that it can make certain ideas thinkable and therefore possible. It is time to develop a new language and to put forward a set of principles, which reflect the reality of arts, and cultural development in Australia and which help us to imagine new possibilities. Below, I put forward some ideas for this which arise from my own practice in community cultural development.[9]

A new language in defence of government support for art

The late Donald Horne wrote and spoke at length against justifying government support for the arts on instrumental grounds, most particularly against the economisation of culture which he described as a "socio economic change…in the conceptions of the economic and what matters in life…it…turns our society into "the economy"" (Horne 2002, 2).

In his notion of "cultural rights" may lay the seeds for the regeneration of a public conversation about the intrinsic value of the arts (Horne 2000, 252). This concept of "cultural rights" marks out the arts and cultural activities as part of the realm of the imagination—and therefore as different from the instrumental—and challenges any elitist connotations ascribed to arts activities by insisting on the right of all to both critically engage with their cultural heritage and new intellectual and artistic production, and to find their own forms of expression.

John Holden also questions the degree to which cultural organisations should be obliged to use instrumental arguments to justify public funding whilst documenting the difficulty of measuring instrumental value due to "complicated and contested assessments of causation" (Holden 2004, 17). He suggests that cultural value should be redefined to recognise art and culture's "intrinsic" value, recognising the subjective nature of cultural experience, practice and identity.

Asserting people's right to art he proposes that a new language, a language of public good and public value which recognises the "moral, creative and collective values expressed by the right to art be developed and applied" (Hewison and Holden 2004, 14).

Removing educational, financial, social and cultural barriers

In making his argument that art should be viewed as a "public good," Holden suggests that public good should be judged by public preference; that is, a public agreement that taxes be used to sustain the arts and culture, to provide educational resources and to give their time to the enjoyment of culture (Hewison and Holden 2004, 13). But how does this stack up in the Australian context?

Australians spend an average of an hour to more than two hours a day doing handiwork and crafts, attending the theatre, concerts, museums and art galleries, performing or making music, listening to CDs, records or tapes or reading books (Australia Council 2002). Almost two million Australians participate in art making activities outside of their normal jobs, but not as hobby activities (Australia Council 2002; Australian Bureau of Statistics 2004). The 2005 National Review of School Music Education undertaken by the Federal Minister for Education received the largest number of submissions—5990—ever received by any Parliamentary Inquiry, indicating the depth of feeling and commitment by many Australian citizens to their children's right to a musical education (Mills 2005).

This high level of engagement in a range of popular, "sub-cultural" and "high" arts activity is not reflected in a widespread engagement with the output of subsidised arts and cultural institutions. Here we see a marked distinction between those who attend, who have attained tertiary levels of education and those who do not attend, who have not (Bennett and Frow 1991; Bennett 1994b; Woolcott 1999).

This distinction is also reflected in significant differences in support for public funding for the arts—and once again the divide is along the lines of educational attainment (Australian Bureau of Statistics 1997).

The message for governments, arts policy makers and arts administrators is clear. Government policies must translate into removing the educational, financial, social and cultural barriers to citizens' active and critical engagement with art and with their cultural heritage.

Recognition and support for individual's rights as cultural producers

Policy needs to acknowledge and resource, where necessary, the right of the Australian people to participate in cultural and intellectual life, not just as consumers, but as active participants in the making and management of our creative and cultural development.

There needs to be an understanding of the necessity of art and how it is a large part of what makes us human; that art is more than individual achievement and that, in our struggle to achieve sustainable communities, art-making is one way of moving beyond the concerns of consumerism and individualism to connect with the "communal" and understand and experience the relationship between past and future.

Recognition and support for the social emphasis of art and culture

Policy responses also need to move into or continue to operate in social spheres. This is not about trying to woo welfare agencies into funding arts projects. Nor is it about the purely instrumental application of the arts to social ends described earlier. This is to do with people's everyday lives; acknowledging that it is not only the artworks produced by artists and the community which are important, but also the capacity of the world of intellect and imagination to create meaning in our lives and to change our habits of thinking, seeing and behaving.

Integrating cultural with social, environmental and economic policy approaches

There is also a need for a clear policy mandate which recognises multiple policy objectives and outcomes, not only arts or cultural objectives and outcomes, and which argues for the development of a cultural framework which illuminates the cultural implications of planning and technical disciplines rather than the establishment of separate cultural policies and "cultural empires" within government.

UNESCO'S 2001 *Agenda 21 for Culture* recognises that the quality of local cultural development depends on the interweaving of cultural and other public policies in urban planning. This Agenda also rejects the treatment of cultural goods and services as mere commodities or consumer goods and emphasises the significance of public spaces in navigating and negotiating meaning, identity and a sense of place.

In some Australian communities in Australia, including the City of Philip in Victoria and the City of Parramatta in NSW, these ideas are well understood and for years there have been attempts to develop and apply these kinds of cultural frameworks rather than distinct cultural policies (Mills 2005).

Recognition and support for cultural diversity

And finally, cultural self-determination should become the *objective* of any national cultural policy, supporting diverse cultures and resourcing diverse forms of cultural expression. "Cultural diversity" should not only describe ethnic diversity, but attempt to describe our country's different cultural heritages and the arts practices arising from them. These are variously described as cultures emerging from particular places (suburban, urban and rural), age groups, racial groups, ethnic groups and interest groups. The fact that people move in and out of a number of these cultures and can belong to more than one at any one time should also be acknowledged.

Policy responses also need to acknowledge and support the interactions that occur between different cultures, and between popular culture and "high" culture.

Who owns cultural policy?

Where can such a policy reside? Who would articulate it? Obviously a national policy needs to reside in government and should be articulated by its politicians and agencies. I have argued in this chapter why the chances of that happening in Australia under the current circumstances are slim.

Perhaps the capacity for policy now lies apart from government; with that array of individuals and organisations with a stake in cultural policy. Perhaps then, we can conceive of a cultural policy framework envisaged and articulated by these communities of interest who may then educate governments on the necessity for such a policy.

References

ATSIC. 1995. *Cultural Policy Framework.* Canberra: Australian Government

—. 1997. *Cultural Industry Strategy*. Canberra: Australian Government

Australian Bureau of Statistics. 1997. *Public Attitudes to the Arts.* Canberra: Australian Government.

Australian Bureau of Statistics. 2004. *Work in Selected Culture and Leisure Activities.* Canberra: Australian Government.

Australia Council for the Arts. 2002. *The Arts: Some Australian Data, Sixth Edition.* Sydney: Australia Council.

—. 2004. *Education and the Arts Strategy 2004–2007.* Sydney: Australia Council.

Australian Labor Party. 2005. *Arts, Culture and Heritage Policy.* Canberra: Australian Labor Party.

Bennett, Tony and John Frow. 1991. An Anatomy of Taste. *Art Galleries: WhoGoes?* Sydney: Australia Council.

Bennett, Tony. 1994a. Research and Cultural Development. In Marcus Breen, ed. *Enhancing Cultural Value: Narrowcasting, Community Media and Cultural Development.* Melbourne: CIRCIT

—. 1994b. Non-Goers: A Social and Cultural Profile. *The Reluctant Museum Visitor*. Sydney: Australia Council

NSW State Government. 1999. *Encouraging the arts in local communities: the Carr Government's cultural development policy.* Sydney: NSW Ministry for the Arts

Commonwealth of Australia. 1975. *Australia Council Act 1975.*

—. 1994. *Creative nation: Commonwealth cultural policy*. Canberra: Department of Communications and the Arts.

Cultural Ministers' Council. 1991. *National Culture/Leisure Industry Statistical Framework*, 4th Edition. Canberra: Government Publishing Service.

Department of Communication and the Arts. 2005. *Orchestras Review.* Canberra: Government Publishing Service.

Dunn, Anne. 2006. *Community Partnerships Scoping Study: Creative Communities*. Sydney: Australia Council

Garrett, Peter. 2007. *New Directions in the Arts: Supporting a vibrant and diverse Australian arts sector*. Australian Labor Party.

Gonski, David. 1997. *Review of Commonwealth Assistance to the Film Industry.* Canberra: Department of Communications and the Arts

Government of Victoria. 1991. *Mapping Our Culture.* Melbourne: Victorian Department for the Arts

Hawkes, Jon. 2001. *The Fourth Pillar of Sustainability.* Melbourne: Common Ground.

Hawkins, Gay. 1993. *From Nimbin to Mardi Gras: Constructing Community Arts*. Sydney: Allen and Unwin.

Hewison, Robert and John Holden. 2004. *The Right to Art: making aspirations reality*. London: DEMOS.

Holden, John. 2004. *Capturing Cultural Value—How culture has become*

a tool of government policy. London: DEMOS
Horne, Donald. 2000. *Into the Open.* Sydney: Harper Collins
—. 2002. The Arts, the Regions and the Economy. *Artwork*: 54. Adelaide: Community Arts Network SA
Kalantzis, Mary and Bill Cope. 1995. *The Paradoxes of Differing.* Sydney: Centre for Workplace Communication and Culture.
Liberal Party of Australia. 1996. *For Arts Sake: A Fair Go,* Arts Policy of the Liberal and National Coalition
Marr, David. 2005. Theatre Under Howard. Philip Parsons Memorial Lecture on the Performing Arts, October 9, 2005, Sydney.
Mills, Deborah. 2001. Culture: a parallel framework. *Artwork*: 50. Adelaide: Community Arts Network SA.
—. 2003. Cultural Planning—Policy Task not Tool. *Artwork:* 55. Adelaide: Community Arts Network SA.
—. 2005. *The Necessity of Art: Claiming Our Right to be Human.* Address to the Local Government Community Services Association of Australia Annual Conference, Melbourne.
—. 2006. *Cultural Development and the Arts in Everyday Life.* Discussion paper prepared for the Australia Council's Scoping Study Reference Group, Australia Council for the Arts.
Mills, Deborah and Paul Brown. 2004. *Art and Wellbeing: a guide.* Sydney: Australia Council
Arts Queensland. 2002a. *Creative Government: Arts and Cultural Activity Across the Queensland Government.* Brisbane: Queensland Government
—. 2002b. *Creative Queensland: the Queensland Government Cultural Policy.* Brisbane: Queensland Government
Regional Arts Australia. 2005. *National Directions for the Arts in Regional Australia: Regional Arts Australia National Community Consultation Report, May – July 2005.* Adelaide: Regional Arts Australia
Saatchi and Saatchi. 2001. *Australians and the Arts*. Sydney: Australia Council
Throsby, David. 2006. Does Australia Need a Cultural Policy? *Platform Papers*: 7. Sydney: Currency House
UNESCO. 2001. *Agenda 21 for Culture.* Available at http://agenda21culture.net (accessed 21 May 2008).
Western Australian Government. 2002. *Focus on the Future: the Western Australian State Sustainability Strategy.*
Woolcott Research. 1999. *Selling the Performing Arts*. Sydney: Australia Council.

Notes

[1] The development and enacting of a satisfactory national cultural policy always seemed unlikely during the Howard Government's reign. For the moment the Rudd Labor Government seems more comfortable sticking to an arts policy rather than venturing into the more difficult territory of 'culture' (Garrett 2007). At least in so doing they are avoiding the tendency by most governments to talk about culture when they mean art, thereby exposing their inability to understand the role that culture plays in everyday life.

[2] The notion of "collaborative art-making" is a critical distinction here and I will explain why later.

[3] This Committee was reinstated as the Community Cultural Development Board in 1990 with the same broad charter as the Committee.

[4] My views on this will not be supported by those adherents of the rationalist and technocratic approach to policy making who believe that policies are based on well reasoned, evidence-based propositions, should be product and outcome driven—rather than process driven—and should discount things which are difficult to measure (see Holden 2004,17).

[5] When the Australia Council commissioned the international advertising firm of Saatchi and Saatchi to conduct research into Australians' support for the arts it foreshadowed the 2001 research findings which framed Australian citizens as 'consumers.' The Council's response to these findings was to fund a range of arts marketing strategies aimed at increasing audiences for subsidised arts activities.

A construction of the public as 'consumer' rather than 'citizen' is inimical to any notion of active participation in intellectual and art production by Australians citizens.

In the light of these divergent value positions, the outrage at the actions of the Australia Council in abolishing the Community Cultural Development Board and locating its new Community Partnerships section within its Audience Development and Marketing Division can perhaps be better understood. From my discussions with the Australia Council Chair and CEO at that time they gave no impression of having been aware of their radical reframing of the Board's activities or of its implications.

[6] Hardly surprising since Regional Arts Australia's *National Directions for the Arts in Regional Australia* report in 2005 and the Scoping Study report were researched and written by the same person, Anne Dunn.

[7] When the New Media Board and Community Cultural Development Boards were abolished by Kemp in 2005 on advice from the Australia Council speculation was rife that the recommendation had been made on prompting from the Minister's office.

[8] The thrust of this section of the policy could be interpreted as being about making high culture available to suburban and rural communities.

[9] The following principles are based on my experience in the community cultural development sector and influenced by the many policy makers and practitioners that I know and have worked with from that field.

CHAPTER FOURTEEN

KNOWLEDGE, POWER AND CULTURAL POLICY: SOCIAL UNDERSTANDING THROUGH COMMUNITY CULTURAL DEVELOPMENT

PAUL BROWN

Introduction

There is a field of social enquiry called "Public Understanding of Science," in which researchers examine the relationship between scientific knowledge and the society it serves. These studies have critiqued the so-called "deficit model" of knowledge production. This model assumes citizens lack the understanding needed for difficult and complex life decisions, but can be brought to understand by the institutions which supposedly generate this knowledge, for example government agencies, large corporations, universities and other centres of research. In opposition to this model, scholars have given focus to the many ways in which expertise lies outside the "official" institutions, with knowledge generated by diverse processes in contexts which are variable and equally valuable. "Lay," "Indigenous" and "Hybrid" knowledges are examples of forms standing strong alongside scientific knowledge (Irwin 1995), even though at the start of the Twenty-First Century, scientific knowledge retains its privileged position in most formal decision making processes.

What has all this to do with cultural policy? This essay reports some of the ways that the creative arts, and particularly the form known in Australia as "community cultural development" (or CCD), constitute a site for the production of important knowledge about complex social, economic and environmental matters. The second part of the argument will explore the transformative power of CCD through examples of "Art and Wellbeing" programs. "Knowledge is Power" is a familiar aphorism argued comprehensively by Enlightenment philosophers. They promoted rationalism and "scientific" inquiry as the means for profound transformations necessary for the betterment of human society. We can argue that the processes of CCD don't just produce knowledge, they allow powerful transformations of the way social and environmental problems are framed, of the way solutions are sought, of the mode of implementation of public policy and of the way society evaluates and monitors the implementation of solutions. In other words, CCD can, and does, transform the whole cycle of policy and decision making.

To elaborate this point, this essay is best read alongside the penultimate chapter in this collection, "Citizen Values and the Arts: New Language for Australian Cultural Policy". There Deborah Mills distinguishes between the instrumental ("arts as a tool") and transformative roles of the arts. This leads her to principles for a national cultural policy that integrates CCD. Seconding this, I will suggest that the last thing we need is a Government sponsored cultural policy which is simply about support for the arts; as if society appends the creative arts for entertainment or as an educative tool for delivering particular objectives. If there's to be Australian Cultural policy or policies, it/they should acknowledge the transformative role of the arts and integrate principles and programs for development with national and local community goals that are cultural, social, economic and ecological.

Understanding and Knowledge

In our 2004 book *Art and Wellbeing*, written as part of an Australia Council study into the arts' contribution to individual and community wellbeing, Deborah Mills and I draw on the concept of culture as a "fourth pillar." Following previous work by commentator Jon Hawkes (2001), we propose that "culture" alongside "economic concerns," "social equity" and "ecological sustainability" have become the elements through which we can all make the great decisions we have to make about difficult and complex life problems.

Another impulse for that research was to explore problem solving through *cross-sectoral* approaches, imperative for government agencies at present.

With much rhetoric about "engaging communities," government acknowledges that the most difficult of problems can't resolve without citizen participation. But what governments can't admit (since it would fly against their interests) is that the structure of modern society now prohibits the institutions of centralised control from *understanding*. Why can we argue that? First, another way of putting it, is that governments lack capacity to produce "socially negotiated knowledge" (knowledge we all agree on), which we need for tackling complex life decisions. There are some good reasons for this. "Evidence-based" processes are deemed important by governments, yet much of the evidence gathering is structured from the centre (through government and industry funding programs), creating a remoteness from the community's sense of what is needed in the context of culturally embedded values and perceptions (Dovers 2005). To use an old phrase, the community doesn't "own" the knowledge. Further, despite the critique referred to above from the field of Public Understanding of Science, the great majority of public participation or engagement processes continue to be structured around the deficit model. We have the situation where governments *think* they understand more than communities, but, structurally, they are unable to. Meanwhile the privileging of scientific and technical knowledge to the exclusion of other forms of knowledge is very damaging to the possibilities for *deeply* understanding the way in which communities need to change in order to solve problems. This isn't a bid for less science in decision-making—for example we can't understand many environmental problems at all without scientific evidence—but it is a bid for science conducted "in public," and in balance with other contextualised inputs.

In a more concrete way, the fact that we live in "low taxing times" also compromises the prospects for our governments to understand new, complex problems and take viable action. In Australia no major political party dare propose higher taxes, though this lack of bravery means a crippled resource base detrimental to effective services. Alongside this we have a late twentieth-century phenomenon of public-private sector partnerships which, complicit in some spectacular recent failures of government infrastructure policy including the 2006 insolvency of Sydney's Cross City Tunnel Consortium and the cost blowouts associated with building Sydney's controversial desalination plant. Such partnerships have the high risk of limiting government action to options commensurate with the commercial interests of large development corporations and investment banks.

These conditions severely constrain the resources available for comprehensive knowledge building and decision making processes—of the "open" type that would admit and rigorously assess the widest possible range of solutions. As Australian governments contend with scarce resources and limited options for action, the hardest-hit of all elements in the policy cycle is the final evaluation and monitoring stage (Dovers 2005)—so that governments lack the capacity to understand successes and mistakes or, from this, to establish appropriate new problem solving programs. For example, some leading policy analysts, including Robyn Eckersley and Stephen Dovers, present comprehensive arguments about how existing structures and norms of centralised government are incapable of delivering outcomes consistent with ecologically sustainable development (Eckersley 2004; Dovers 2005).

Perhaps this lack of understanding is the destiny of democracy. Democracy, with its promises in the later part of the twentieth century of greater public participation, co-evolved with complexity and worsening technological and social hazards, all makings of the Industrial Revolution and accelerating globalisation. As the dangers and complexities of life increased, so did our desire to manage them. And manage them we did, more or less, until recently when dangers such as nuclear threat or human induced global warming have seemed beyond the reach of central systems of control. Characterisations like these caused the sociologist Ulrich Beck to label ours a "Risk Society" (1992), pointing out that runaway globalised "bads" boomerang back to affect us all, regardless of our socioeconomic status or location on the planet.

Many social theorists see the only hope in this to be revitalised "civil" or "grass-roots" movements, capable of taking the action now seemingly out of reach of centralised government. Beck and his colleagues write of an "individualisation" and a "sub-politics" by which the responsibility for taking decisions moves out, away from centralised government, into the hands of communities and networks, for example, through local organisations and participatory processes. As an accompanying trend, we can also note that the project of *globalization* is fractured and incomplete (Sachs 1992). It is met by a *localisation* that remains rebellious and individualistic, and which underscores the importance of *context* in decision making.

All this adds up to a proposition: if we are trying to find those places where knowledge can be generated and negotiated, and where good decisions can be taken, we need to visit what is "local," "individual," "civil," "grass roots," and "rebellious." This is where citizens are trying *to make meaning out of life*, the essential human endeavour which Donald Horne (among others) equated with *culture* (Headon, Hooton and Horne

1995). But in desperate times, meaning isn't enough—communities need to generate working, contextualised knowledge which, in the absence of government understanding, can serve the needs of critical decision making processes. How could this be done?

Trust, Knowledge and CCD

One of the missing ingredients in attempts by governments to centrally generate socially agreed knowledge is trust. Trust is important for "knowing what to do" since, in its various forms (individual, group, organisational), trust is a pre-condition for social agreements about what passes for useable knowledge. The best explanation of how this works comes from the Australian social theorist Gavan McDonell (1997). In trying to understand the "glue" that allows adversaries to enter into debate (the starting point of knowledge making), McDonell delineates a form of *provisional* trust that he calls "suspended doubt". This is the situation that exists when people, two people even, are prepared to sit around a table and set aside, temporarily, their greatest differences in order to progress toward some common goal. The act of suspending doubt opens up a space, and opens up time for unease and dissent to play out, and for knowledge to be constructed. McDonell goes on to argue that trust (especially the provisional form) is of greater importance than rationality in decision-making processes.

McDonell's observations derive from his experience in formal (government sponsored) public participation associated with environmental problems. However, in the current debate about cultural policy, we might equally focus our attention on the conduct of creative arts projects and programs. When these use CCD approaches, it is easy to identify the same suspension of doubt, the same provisional trust that McDonell describes. Community theatre projects, for example, thrive on communicative processes that are engendered in an environment of suspended doubt. This environment is created when diverse members of a community enter into a time and a place (a workshop, a rehearsal) in order to make a project (Brown and Crittenden 2007).

The next section will explain further how this all works, and give some examples. But first we need to think through the idea of "community" in more detail. Having hinted that there is diversity in any community, we can go further (drawing on community development literature) and point out the possibility that there's no such thing as "community" only the *desire* for community (Brent 2004). Each time we try to delineate a community—whether by geographic boundary, or by function, or by common political interest, or in some other way—we find the group's members have such different backgrounds, and that there is such internal unease and fracturing, that we must hesitate to speak of "cohesion," and "unity," and "identity."

In this essay, "community" can continue to stand as a meaningful if contested term. What's important for the discussion is that the very diversity, unease, divisions and conflict in communities make CCD a knowledge-making process, since the trust CCD generates provides the circumstances for a deep exploration of both evidence and values, albeit in a state of suspended doubt. Because people come with different baggage they will contest ideas, sparks will fly, and values will be up for grabs. But in a trusting environment, participants will manage their dissent, and this "managed dissent" is the catalyst for knowledge. There may be no permanent cohesion in a community, but there *can* arise cohesive streams of thought, knowledge and ultimately action.

Community Cultural Development: Programs of Art and Wellbeing

CCD is a collective process; participatory in terms of program initiation and steerage and in the conduct of creative activities. Professional artists work in partnership with communities in a relationship that breaks down any sense of the "expert providing for the amateur," substituting instead shared responsibility, and a validation of the values, skills and experience brought by all participants. Many programs include government and industry decision makers directly in creative activities and/or build in communicative structures that "deliver messages" into the heart of decision making processes. (For in-depth case studies of CCD programs and projects that demonstrate these outcomes see Mills and Brown 2004.)

In its practice, CCD involves a wide range of art forms: from performance to visual arts, from film and video to writing, oral history and storytelling. Its creative outcomes may be everything from public art to festivals, theatre and dance performances, exhibitions, publications and seminars. All of these activities—and there are many others I could mention—are characterised by the collaborative and empowering processes by which participants engage with creative activity. As CCD practice has developed in Australia, communities and artists across the country have collaborated on a wide range of projects and programs. These have involved health centres, multicultural organisations, prisons, public housing agencies, environmentalists, educationalists, trade unions, local governments, urban planners, youth centres and women's groups: communities eager to use a range of creative processes, to develop skills and express their concerns and aspirations to a wider audience (Mills and Brown 2004, 6).

The cases studies I collected with Deborah Mills during 2004 secure the connections between arts practice (in particular CCD) and policy arenas of health, ecologically sustainable development, public housing and place, rural revitalisation, community strengthening, active citizenship, social inclusion, and cultural diversity. Here are some of the examples:

1. Big hART is a multi-artform organisation working with people marginalised in rural, regional or other isolated contexts. Domestic violence, suicide, youth crime, health and housing are areas of concern addressed in their projects. Since 2002, in partnership with residents, community workers, artists, a major theatre company (Sydney's Company B), and government agencies, Big hART devised a range of artworks at the Northcott Public Housing Estate in Sydney's inner city. This program has used oral history, and visual and performing arts—painting, film, music, dance, theatre—and aimed to transform government housing policy and decision making structures, while providing a focus for self-determined solutions to community issues such as loneliness, suicide, violence, and crime (Mills and Brown 2004, 25-28).

2. When the theatre company Somebody's Daughter devises scripts and makes performances, they make and communicate new knowledge about the lives and aspirations of women prisoners. The company's participatory processes involve workshops inside and outside prison, intensive long term activities with youth at risk, script writing by prisoners and ex-prisoners, shows inside prisons, publication of participants' writing, public discussions, tours, media coverage, and website information. An important aim is to educate media and the public about the reality of prison, and to break down stereotypical views of prisoners so that policy making can reflect a sophisticated understanding. The inclusion of prison staff and government officials in project groups is an important communicative strategy (Osmotherly 2002; Mills and Brown 2004, 22-24).

3. The 2002 Adelaide Festival saw the integration of CCD principles into one of Australia's most prestigious arts festivals. In a process steered by artsworkers and researchers, the Festival planned a suite of projects concerning Maralinga, the South Australian site of British Nuclear Testing in the 1950s. One of these was a paintings and design project at Oak Valley, a remote Maralinga Tjarutja community, where—through mentoring, skills development and the painting of acryllic artworks depicting the bombs and their aftermath—the community generated and communicated new knowledge about the long term impacts of nuclear testing on people, animals and the land (Page and Wallworth 2003). It followed another successful arts project in which Maralinga Tjarutja painted their relationship with their country (including birthplaces), producing artworks that became documentary evidence in land rights determinations (Allerton 2001).

Since the Festival, another of the Maralinga projects has taken shape, as an oral history and verbatim theatre project involving nuclear veterans, artsworkers, and academics in Britain and Australia. This project is providing contextualised knowledge about veterans' physical and mental health, and communicating this knowledge through performance, at a time when veterans are negotiating for compensation and recognition (Brown 2006). All these projects serve to strengthen advocacy and networking within communities, developing their capacity to undertake long-term development programs.

4. The Atherton Tablelands region in North Queensland was designated within the Australian Government's "Sustainable Regions" program. Here regional economic development relies on enhanced leadership skills, linkages between cultural and other sectors, growth of new industries—in some cases replacing industries no longer viable—and learning for sustainability across the region. Strategies include public art projects, a youth entrepreneurial program (including new media industries), an Indigenous arts project linked to housing development, and trade "expos" that use arts and craft industries to lead new pathways for international export from the region. In this case, the wide range of cultural projects is seen as a key learning strategy, in which knowledge about regional characteristics and patterns of development inform decision making across many sectors (Mills and Brown 2004, 72-73).

5. "Small Towns Big Picture" is the name given to a long range research and development project aimed at revitalisation of five towns in

rural Victoria. Like the Atherton Tablelands project, here cultural activities are not only about creative outputs, but become the engine room for debate, research, and knowledge building about viable development strategies. The project grew from a need for analysis and strategic planning, including "Progress Indicators" across a "Triple Bottom Line" of social, economic and ecological dimensions. Linking a research centre and Victoria's Cultural Development Network, the program involved eight community artists working alongside researchers, conducting workshops across various artforms, which engaged citizens effectively, where previous approaches had failed (Rogers and Ryan 2001; Mills and Brown 2004, 91-93).

6. SunRISE 21 is an organisation based in Mildura on the Murray River established to conduct planning processes for sustainable regional development. It initiated an "Artists in Industry" program addressing the viability of the Sunraysia region's farming community and irrigation practices, which brought together six creative artists (sculpture, theatre, music, new media and photography) with community groups and those institutions which are central to natural resource management and economic development in the region. They include regional and national research organisations, a regional arts centre, peak bodies for irrigation and horticulture and a consortium concerned with salinity management. A key objective has been to use creative processes to engage the interests and challenge the thinking of community and industry networks, in a situation where the community has strongly divergent views about the meaning and practice of sustainability. Networks and business linkages have been strengthened, while debate and learning has been achieved. The various arts projects have increased critical engagement and positive attitudinal change amongst individuals and stakeholder participants, and caused re-evaluation of sustainability questions. Scientists involved in the program have participated in and benefited from the building of cross-sectoral knowledge (Vivian 2000; Mills and Brown 2004, 38-40).

7. Turning to a particular—and older—example of theatre as knowledge-building, "Murray River Story" was a 1988 project of the Murray River Performing group (now Hothouse Theatre). Concerning the ecological problems of the Murray Darling Basin (for example, salinity, over-use of land, erosion, species extinction, chemical and water pollution) it stands as a relevant example of participatory community theatre involving artists, farmers, environmental activists, public servants, recreational clubs, scientists and decision makers. It helped catalyse

community debate about river problems at a critical time—when new governance structures were mooted for the Murray Darling Basin. Through CCD processes, the project allowed "negotiation" of meanings, language and actions amongst diverse participants, with the large scale outdoor performance involving 200 people representing an "agreed" understanding of problems and possible solutions (Brown and Crittenden 2007).

One participant in the project, also an influential decision maker, was Terry Hillman, Director of the Cooperative Research Centre for Freshwater Ecology, an institution responsible for research and advice on the Murray Darling Basin. In assessing the project fifteen years after the performances, he reflected that community theatre helps overcome the difficulties in feeding the needs and desires of communities into river management arenas, and that it gives balance to universal scientific knowledge through production of local, contextualised knowledge:

> There needs to be an opportunity in the process of knowledge building to allow individuals to question the safety of the reliance on scientific knowledge. Theatre can allow the public to raise these questions and challenge these systems. (Hillman 2003, 2)

In *Art and Wellbeing* we were prompted by such ideas to explore how CCD processes could sit alongside planning systems now in place for the Murray Darling Basin, in search of the "deep clues" about seemingly intractable land and water management problems. The current Integrated Catchment Management Plan identifies responsible institutions and agencies across all tiers of government, industry groups, landholders, community organisations, and peak bodies. For all these agencies, CCD offers research-based participatory processes that could develop and negotiate common understandings of natural resource management issues. At stake is the need for essential changes at social, cultural, technical and economic levels, which will involve re-negotiation of values and management systems (Mills and Brown 2004, Appendix 2).

All the above examples show cultural activity orchestrated to foster trust, and from this knowledge, about pressing social, environmental and economic problems. What does this body of work, and generally the Australian experience of CCD, mean for Australian cultural policy?

Commentary: Where to with Cultural Policy?

In the final section of this essay, I won't try to lay out in full the arguments for integrating cultural policy with other policy areas; Deborah Mills argues this convincingly in the previous chapter. With regard to centralised government, she concludes that cultural policy must begin with a sensitivity and awareness of the cultural consequences of *all* government policies and activities, and that the work of a cultural policy should be to establish a cultural sensibility in the way governments think and act. It should not, she says, be about the "establishment of separate cultural policies and 'cultural empires' within government" (Mills, this volume).

She argues that, while government has its place (meaning all three tiers of government in Australia), there is an array of communities of interest, made up of organisations and individuals, with a stake in cultural policy (Mills, this volume).

Implications of this breadth of governance structures responsible for cultural policy is the theme I now want to take up. As argued earlier in this essay, any attempt to delineate cultural policy must first take account of the "sub-politics" of modern society, which sees the decentralisation and individualisation of knowledge production and decision making. All the Art and Wellbeing programs and projects outlined above show communities developing streams of thought and action around their self-determined priorities, with the enactment of solutions involving, though not dependent on, centralised government. Such approaches help realign the meaning and practice of democratic governance.

What we observe about CCD is consistent with studies of participatory practice across many fields. In Australia, some of the most interesting commentary arises from the "Business for Sustainability" movement and other environmentalist fields concerned with organisational change. Here new political theory and emergent management theory find common ground in debates about new governance styles and structures. For example, in exploring the response of organisations (community, business, government) to sustainability imperatives, Suzanne Benn, Dexter Dunphy and Andrew Martin characterise modern democracy and governance as follows:

> [1.] Community-based networks involving multiple stakeholders link different types of knowledge and facilitate knowledge development and diffusion.

[2.] Defamiliarising narratives and storytelling can develop a shared "ecocentric" understanding across organizations.
[3.] "Feminine" collaboration leadership styles support diverse understandings of values, knowledge, experience and opinions.
[4.] Non-hierarchical networks can support a multiplicity of meanings yet allow ongoing collaboration.
[5.] Open and critical debate can increase awareness and the political efficacy of all participants.
[6.] Reflexivity can be fostered through engaging in extra organizational tasks.
[7.] Governance is decentralized, reflexive, diversity and flexible, and based on high frequency of communication. (Benn et al 2008)

The implications for cultural policy are significant. To argue this, we can first summarise what has previously passed for cultural policy in Australia. Historically, approaches have emphasised the intrinsic value of creativity and ways to foster what is artistically "new"; then more recently there has been a drive towards arts-industry development under the banner of "creative industries".

In some non-arts sectors, government policy makers have also given emphasis to the role of creativity as the primary attribute underpinning innovation and progress, or the role of the arts as an educative "tool". This brings the creative arts into focus for their instrumental "use value" to industry and government searching for new catalysts and delivery mechanisms. Such approaches throw up particular policy questions, for example: 1. "How can the artists among us be granted the most stimulating and supportive environment in which to bring forth their creativity?" (a question assuming "intrinsic" value of the arts), or 2. "How can we secure the business position of the arts as a viable and growing industry in the context of 'new global economy'?" (creative industries), or 3. "How can the arts be used more effectively as an input to activity in other sectors?" ("use" value).

Important though they are, none of these questions is adequate to the challenges for cultural policy implied in Benn, Dunphy and Martin's characterisation of modern democracy and governance. In developing diverse and integrated cultural policies, we should be asking: "How can creative activities facilitate trust, knowledge production and communication between organisations and across community networks?". And if organisational change is our objective, we should be asking: 4. "How can the creative arts help us understand the positions and underlying values of stakeholders in conflict?", since without that understanding, negotiated and agreed reforms will elude us. Ultimately, in response to ecological and social crisis, we should also ask: 5. "How can the creative arts infuse decision making processes with new language for debating and salvaging the relationship between humans and the rest of nature?"

Now we are again sailing comfortably close to some of Jon Hawkes' ideas about culture—outlined earlier—as the "Fourth Pillar" (2001). Also, there are approaches to cultural policy at hand which do address these questions in large measure. For example, a "creative society" or "social capital" approach would assume that individual and community capacity-building arises through creative arts processes. The arts become a means for developing personal and collective attributes such as leadership, negotiation, communication, and problem solving. If possessed by communities, these attributes can effect solutions to otherwise intractable social, economic and environmental problems.

Our Australia Council research has pointed out the successes of CCD as an approach which goes beyond "creative industries" or "use value" approaches to develop social capital and ultimately a "Creative Society". In *Art and Wellbeing* we discussed the way the strong ties *within* communities ("bonding social capital") rely on:

> a sense of personal and collective trust and the development of shared values within groups. Such outcomes are cornerstones of community cultural development, just as they are seen as important pre-conditions for policy and action in non-arts sectors. As community knowledge and trust grow, people discover new solutions and establish a shared commitment to those values, goals and means of achieving them. These improved participatory practices can enhance the capacity for change and resolve and/or avoid conflicts likely to arise in the promotion of policy changes. (Mills and Brown 2004, 10)

Such outcomes sit well alongside Benn, Dunphy and Martin's characterisation of democracy and decision making *within* organisations and networks. Likewise the importance they place on communicative and shared understanding *between* social structures is addressed through the increasing appearance in CCD of strategies for enhancing "bridging social capital", which is seen by some as the glue which connects between diverse community groups (Flowers and McEwen 2003). In *Art and Wellbeing*, we identified more specialised bridging strategies in many of the case studies, strategies which "send messages":

> between localised groups and expert systems, such as those systems utilised by agencies charged with managing wellbeing issues and informed by expert knowledge. Projects which originate at the local level can achieve significant advances in wellbeing for individuals and communities. But they can also have important influences in state and/or federal decision-making fora, through deliberate strategies that make links between communities, and which transport participants, ideas and creative outputs into the decision making realm of government agencies. This can resolve the tension that exists between centralised decision making processes and bottom-up processes of community based decision making. (Mills and Brown 2004, 10)

Conclusion

In the deficit model of decision making, citizens are seen to lack understanding and are in need of education. Under the great project of "modernity" most government decision making assumed a deficit of knowledge amongst citizens, assuring a "decide and defend" approach which seeks to "correct" the views of dissenters by providing them with knowledge. We live now with the reverse of that situation. Citizens, armed with contextualised lay knowledge, are able to address a deficit of socially negotiated knowledge inside government.

CCD, with its benefits for making links, sending messages, engendering trust and making knowledge, helps resolve the problematic relationship between diverse governance structures and centralised government. In integrated and democratised cultural policy, support for CCD will vitalise the way decision makers respond to some of Australia's most complex challenges. Business, government and community organisations can become involved in community cultural development as a means of building trust, knowledge and social capital as preconditions for joint decision making about complex issues.

Acknowledgements

A version of this essay was presented for the Byron Bay Writers Festival, August 2005. With regard to the title, I acknowledge ideas generated by staff and students in a University of New South Wales postgraduate course called "Knowledge Power and Public Policy". I thank my colleague Deborah Mills for numerous discussions. Our research for the Australia Council, published under the title *Art and Wellbeing*, underpins this essay.

References

Allerton, Louise. 2001. *Pila Nguru: Art and Song from the Spinifex People*. Spinifex Arts Project.

Beck, Ulrich. 1992. *Risk Society: towards a new modernity*. London: Sage.

Benn, Suzanne, Dexter Dunphy, and Andrew Martin. 2008. Governance of Environmental Risk: new approaches to managing stakeholder involvement. *Journal of Environmental Management*, in press.

Brent, Jeremy. 2004., The desire for community: Illusion, confusion, paradox. *Community Development Journal,* 39, 3:213-223.

Brown, Paul. 2006. Maralinga: Theatre from a Place of War. In *Unstable Ground: the Politics of Place and Performance*, ed. Gay McAuley. Berne, Peter Lang.

Brown, Paul, and Xanthe-Rose Crittenden. 2007. Nature Moves Centre Stage: Ecological Knowledge in Community Theatre. *About Performance*, 7.

Dovers, Stephen. 2005. *Environment and Sustainability Policy: Creation, Implementation, Evaluation*. Sydney: Federation Press.

Eckersley, Robyn. 2004. *The Green State: Rethinking Democracy and Sovereignty*. Cambridge Massachusetts: The MIT Press.

Flowers, Rick and Celina McEwen. 2003. *The Impact of 'Reigniting the Communnicty' and 'The Torch' on community capacity building*. Sydney: Centre for Popular Education.

Hawkes, Jon. 2001. *The Fourth Pillar of Sustainability: Culture's essential role in public planning*. Melbourne: Common Ground.

Headon, David, Joy Hooton, and Donald Horne (eds.). 1995. *The Abundant Culture: Meaning and Significance in Everyday Australia*. Sydney: Allen and Unwin.

Hillman, Terry. 2003. Interviewed by Xanthe-Rose Crittenden, Albury, 21 May.

Irwin, Alan. 1995. *Citizen Science: a study of people, expertise and sustainable development*. London: Routledge.

McDonell, Gavan. 1997. Scientific and everyday knowledge: trust and the politics of environmental initiatives. *Social Studies of Science*, 27: 834-835.

Mills, Deborah and Paul Brown. 2004. *Art and Wellbeing: a guide to the connections between community cultural development and health, ecologically sustainable development, public housing and place, rural revitalisation, community strengthening, active citizenship, social inclusion and cultural diversity*. Sydney: Australia Council.

Osmotherly, Jan, 2002. *Somebody's Theatre Company Evaluation Report*.

Melbourne: Osfield Consultants.

Page, Alison and Lynette Wallworth. 2003. Placemaking,/Wellbeing and the Adelaide Festival of Arts. Conference Paper, Arts and Ecology Symposium, School of HPS, University of NSW.

Rogers, Maureen and Roberta Ryan. 2001. The Triple Bottom Line for Sustainable Community Development. *Local Environment*, 6, 3: 279-289.

Sachs, Wolfgang, ed. 1992. *The Development Dictionary: A Guide to Knowledge as Power*. London: Zed Books.

Vivian, Helen. 2000. *Interceptions: Art, Science and Land in Sunraysia.* Mildura Arts Centre and Artmoves Inc.

CONTRIBUTORS

Lisa Andersen

Lisa Andersen is Manager and Senior Researcher on the Australian Research Council funded CAMRA Project—Cultural Asset Mapping for Planning and Development in Regional Australia—and is also Community Engagement Coordinator for UTS Shopfront Community Program at the University of Technology, Sydney. Her research interests include audience and market development, regional cultural industries, community cultural development, and community-engaged research. Recent industry reports include: Quality of Light, Quality of Life: Professional Artists and Cultural Industries in and around Broken Hill (2007), Common Ground: Cultural Festivals in NSW's Northern Rivers Region (2006), and 'Archie in the Country': The Archibald Prize Exhibition at Cowra Art Gallery (2005). In 2005 she was co-winner of the Australian Award for University Teaching for the provision of education services to the community.

From 2004, with Jill Eddington, Lisa developed and co-hosted a series of forums as part of the Byron Bay Writers Festival where leading thinkers considered future public policy directions for arts and cultural industries. This book came out of those forums.

Anne Bamford

Professor Anne Bamford is Director of the Engine Room at Wimbledon College of Art, University of the Arts, London. She has been recognised nationally and internationally for her research in emerging literacies and visual communication and completed the UNESCO global impact of the arts in education research compendium. Through Anne's research she has pursued issues of innovation, assessment, social impact and equity and diversity. She has also researched extensively cyber bullying and its effect on adolescent identities. In 2002 Anne was awarded the Australian Institute for Educational Research, Outstanding Educational Research Award.

John C. Barsness

John Barsness is Executive Director of Montana Arts, a statewide arts organisation based in Bozeman, Montana, USA, and a former Associate Professor of Art at Edgewood College, Madison, Wisconsin. He also served on the President's Advisory Committee on the Arts, Washington, DC, from 1995 to 2001, and is currently a member of Americans for the Arts' State Arts Advocacy Council.

Pamille Berg (AO)

Pamille Berg is an award winning Public Art Consultant specialising in major public art strategies and master plans. Among other major projects, she completed the Public Art Master Plans for the Parliament House Art Program, the island of Maui (Hawaii), the Australian National Botanic Gardens and St. Patrick's Catholic Cathedral in Parramatta. In 2004 Pamille was appointed an Officer in the Order of Australia (AO) for her services to public art, with particular note of her creation of opportunities for emerging artists and Indigenous artists in remote communities.

Eva Cox

Eva Cox was born Eva Hauser in Vienna in 1938, and was soon declared stateless by Hitler. So she grew up as a refugee in England, till 1946, and arrived in Australia aged ten. These early experiences influenced her rejection of injustices. She has been an active and irrepressible advocate for creating more civil societies. She is a long-term member of the Women's Electoral Lobby and an unabashed feminist.

Eva until recently was lecturing in Social Inquiry at the University of Technology, Sydney. She is a sociologist by trade and has undertaken many research projects and has published widely and eclectically in books, journals and newspapers. Her current work builds on her Boyer Lectures (1995) on social capital and her commitment to making societies more civil. One current interest is encouraging NGOs to be ethical and advocate effectively. Another is monitoring the effects on sole parents of welfare to work policies.

Paul Brown

Dr Paul Brown co-ordinates an Environmental Policy Program for the Faculty of Arts and Social Sciences at University of New South Wales. Based in the School of History and Philosophy, his research interests span public participation, community-industry relationships, and history and sociology of technology. He is also a script writer for theatre and film, and an arts practitioner with experience in community cultural development. He was co-founder of Sydney's Urban Theatre Projects and the author of "Aftershocks" a verbatim play about the Newcastle earthquake. He is the facilitator and script editor for the "Half a Life" project concerning the experience of Maralinga nuclear test veterans and their families.

Chris Gibson

Chris Gibson is Associate Professor in Human Geography at the University of Wollongong. His research interests include: cultural industries and regional economic development strategies, popular music, tourism and festivals. Recent books that he has authored or co-authored include *Music and Tourism* (2005), *Deadly Sounds, Deadly Places: Contemporary Aboriginal Music in Australia* (2004), and *Sound Tracks: Popular Music, Identity and Place* (2003). He is the co-author of the forthcoming *Cultural Economy of the Popular Music Industry* (Edward Elgar, 2009).

Cathy Henkel

Cathy Henkel is an award-winning documentary filmmaker who lives in the Northern Rivers region of NSW, Australia. Her screen industry credits include *The Burning Season*, *The Life and Legacy of Spike Milligan* and *The Man who Stole my Mother's Face* (winner of Best Feature Documentary at the Tribeca Film Festival in New York and the IF Award for Best Documentary in Australia.) Her work has been screened in 25 countries and is delivered on broadcast television, online and on DVD. She is currently a PhD candidate at Queensland University of Technology examining the prospects of screen and creative industries in the Northern Rivers region.

Christopher Madden

Christopher Madden is Research Analyst at the International Federation of Arts Councils and Culture Agencies. He has worked for a variety of arts and culture agencies in Australasia, including the New Zealand Ministry of Cultural Affairs, Creative New Zealand, the Australia Council for the Arts, and the National Centre for Culture and Recreation Statistics of the Australian Bureau of Statistics. He has extensive experience in modelling the cultural sector, developing frameworks for cultural statistics and cultural indicators, and in analysing data for cultural policy. A significant part of his research has addressed issues of evidence-based advocacy.

Deborah Mills

Deborah Mills has a diverse background in community and cultural development and a strong record in public sector social and cultural policy development in Australian commonwealth, state and local government spheres. She worked for the Community Cultural Development Board of the Australia Council and its predecessor the Community Arts Board for thirteen years, the last five as Director of the Board. Deborah is the Director of Deborah Mills & Associates, a cultural planning and development company.

Deborah was co-author, with Dr Paul Brown, of Art *and Wellbeing,* an Australia Council publication which examined the connection between community cultural development and government wellbeing initiatives.

Tony Moore

Tony Moore is a cultural historian and author who recently completed a PhD on Australian bohemia at the University of Sydney. He was Commissioning Editor of Pluto Press from 1997 to 2008 and a documentary maker at the ABC from 1988 to 1997. Tony's most recent book is *The Barry McKenzie Movies* (Currency Press) and he was awarded the NSW History Fellowship in 2007 to write a history of political prisoners transported to Australia in the convict period. Tony was a member of the ABC National Advisory Council, is a Fellow of the Centre for Policy Development and sits on the executive of the NSW Fabians.

Martin Mulligan

Martin Mulligan is a Senior Research Fellow in the Globalism Institute at RMIT University in Melbourne where he specialises in the sustainability of local communities in the context of globalisation. He was project manager for the Institute's three-year study of arts, celebrations and community wellbeing conducted for VicHealth. He was also the initiator and coordinator of the Two Fires Festival of Arts and Activism held in Braidwood, NSW, in March 2005, which sought to celebrate and extend the legacy of Judith Wright. He was co-author with Stuart Hill of the book *Ecological Pioneers: A Social History of Australian Ecological Thought and Action* (Cambridge University Press, Melbourne, 2001).

Kate Oakley

Kate Oakley is a writer and policy analyst, specialising in the cultural industries, cultural policy and regional development. She is a Visiting Professor at the Department of Cultural Policy and Management, City University, in London.

She was the author of the *Creative London* Report in 2004 and has worked with a number of cities and regions, both in the UK and internationally, on public policy in the cultural industries. Recent publications in this field include *London's Creative Economy: An Accidental Success?* (2007) co-written with John Knell, published by the Work Foundation, and *Better Than Working For a Living?* (2007) a study of labour markets in festivals and events, published by City University.

Kate's current research interests concern labour in the cultural industries, urban policy and art education.

Jeremy Sim

Currently a bank consultant based in Singapore and Beijing, Jeremy Sim was formerly the lead project manager for a European Union initiative to stimulate innovation and creativity in northern Europe's industries and universities. He has advised and presented on creative industries development models, especially those adopted in Scotland and Sweden, to various authorities and conferences in Asia, Australia, Europe and North America.

David Throsby

Professor David Throsby is one of the world's leading cultural economists. He is Professor of Economics at Maquarie University, Sydney, and is currently Chair of the NSW Arts Advisory Council and member of the Experts Committee for drafting the UN Convention on Cultural Diversity (2003-04) UNESCO. David has been a consultant to the World Bank, the OECD, FAO and UNESCO, as well as many government organizations. His book, *Economics and Culture* (Cambridge University Press, 2001) has been translated into five languages. In 2008 David took part in the Prime Minister's Australia 2020 Summit and the Towards a Creative Australia panel.

INDEX